JAGUAR
XJ-S

The Full Story of Jaguar's Grand Tourer

First published in 2000 by Veloce Publishing Plc., 33, Trinity Street, Dorchester DT1 1TT, England. Fax: 01305 268864/e-mail: veloce@veloce.co.uk/website: www.veloce.co.uk

ISBN: 1-901295-43-5/UPC: 36847-00143-8

British Library Cataloguing in Publication Data -
A catalogue record for this book is available from the British Library.

Typesetting (Bookman), design and page make-up all by Veloce on AppleMac.
Printed and bound in the UK.

JAGUAR XJ-S

The Full Story of Jaguar's Grand Tourer

BRIAN LONG

VELOCE PUBLISHING PLC
PUBLISHERS OF FINE AUTOMOTIVE BOOKS

CONTENTS

To Waylon - Jaguar fan and an inspiration.

INTRODUCTION &
ACKNOWLEDGEMENTS

This is the full story of Jaguar's Grand Tourer, a much-admired yet misunderstood model that took many years to gain the recognition it deserved. Most thought it was a direct replacement for the E-type, and judged it as such. In fact, for numerous reasons, the intention had always been to move the new V12 supercar into a different market sector to Coventry's most famous sports car. The timing of the launch could have been better too: British Leyland's problems and an era blighted by a series of fuel crisis highlighting how fuel-thirsty the early cars were did not help the new Jaguar.

Fortunately, the model wasn't allowed to die, and the HE version of 1981 brought the car worldwide critical acclaim. As the years went by, a six-cylinder XJ-S became available, along with a Cabriolet and a Convertible. Quality had improved in the meantime, ensuring the XJ-S kept its position in the top division of luxury sporting machinery, despite an ageing concept.

During 1991, the car was treated to a major facelift, followed by another one a couple of years later to keep it competitive. By the time production ended, over 115,000 cars had been built; a very impressive figure for a vehicle of this sort. Along the way, the XJ-S also inspired a great many 'specials' and acquired an excellent record in top-class competition.

I would like to take this opportunity to thank the numerous members of staff at Jaguar who have helped me over the years. When writing this book I'd already been a regular visitor to Browns Lane for more than a decade, during which time I'd been lucky enough to speak to many of the key personnel behind the cars described in this book. Thank you, everyone!

I would particularly like to acknowledge the help of someone who, for me, is the world's foremost motoring historian - Paul Skilleter. Paul was good enough to share his archive material with me, and has taught me a lot over the years. I'm very proud to be able to call him a friend.

Brian Long
Chiba, Japan

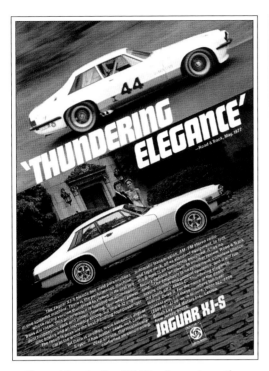

Two sides to the XJ-S's character - the noise and brutal power of the Group 44 racer, which brought a great deal of success for Bob Tullius' Virginia-based concern, and the supreme elegance of the S in its role as a Grand Tourer.

One of the most famous of all Jaguar adverts, from the time of the XJ-S launch. The headline implies that the likes of Ferrari, Maserati, Lamborghini, De Tomaso, Mercedes and Porsche should start to worry. The Turin link is not so much an attack on Fiat and Lancia (with the passing of the latter's Stratos, Lancia had nothing for the road to compete in the supercar class), but the fact that the Fiat Group owned a massive stake in Ferrari.

Left: American advertising for the XJ-S dating from 1976. The Americans often referred to the model as the S-type, even in official paperwork issued from the company's US headquarters in New Jersey.

The 1983 HE model in US trim.

PRESENTING JAGUAR FOR 1979.
THE LATEST AND THE BEST IN A LONG AND ARISTOCRATIC LINE.

he 1979 Jaguars are the finest expression to ate of our uncommon and uncompromis- g approach to building great luxury cars. The luxury is deep and so well thought-out nat the Jaguars come to you complete: there re no factory options whatever. Real leath- r seats, thermostatically-controlled heating nd cooling, stereo AM/FM radio and tape ystem: it's all there, adding to the notable alue inherent in all Jaguar cars. But, as ever, the finest expression of any aguar motorcar is the way it moves. The 1979 Jaguars have the incredible re- ponse of electronic fuel injection teamed vith either the double overhead-cam Six in

the XJ6 sedan, or the aluminum V-12 in the XJ12 sedan and the S-type. And Jaguars have the surefootedness of four-wheel indepen- dent suspension and power disc brakes on all four wheels. They have power-assisted rack and pinion steering and an overall feel for the road that is rare, indeed, in luxury cars. Another uncommon feature of the 1979 Jaguar is its warranty: for 12 months, regard- less of mileage, your Jaguar dealer will re- place or repair any part of the car that is defective or that simply wears out, provided only that the car is properly maintained. The only exceptions are the tires, which are war- ranted by the tire manufacturer, and the

spark plugs and filters, which are routine replacement items. Even then, if the plugs or the filters are defective, Jaguar will replace them. There has never been a better time for you to own a Jaguar, for the simple reason that there has never been a better Jaguar than our 1979 edition. For the name of the Jaguar dealer near- est you, call these numbers toll- free: (800) 447-4700, or in Illi- nois, (800) 322-4400.

BRITISH LEYLAND MOTORS INC., LEONIA, NEW JERSEY 07605.

American advertising for the 1979 model year. Note the changes needed to make the Series 2 saloon meet Federal regulations.

The TWR Jaguar XJ-S in its 1983 livery.

The TWR racer in its final form, pictured recently at Browns Lane; the exhaust noise from the V12 is glorious.

A 1937 SS100 with the newly-introduced 3.6 litre XJ-SC - the first official open Jaguar to reach the market in almost a decade. The photograph was taken at Wappenbury Hall, Sir William Lyons' home from 1937.

The Daimler-S - an attempt to break into a different market with the XJ-S. It was appraised in an American dealer's clinic, but was rejected, mainly because of the lack of knowledge about the Daimler name in the States. Just one Daimler prototype was built.

Launched in 1983, this is a 1987 model Lynx Eventer estate based on the XJ-S. Even after the S was restyled, the Lynx concern continued to produce an equivalent to this handsome carriage. Lynx also carried out a number of other conversions of the Coventry car.

Tasteful British advertising from late 1987.

A Hess & Eisenhardt XJ-S Convertible, sadly, produced for the US market only. It was available from 1986 to 1988, in which time around 2000 were built.

One of three XJ41/42 prototypes built by Karmann in 1989. Two had the targa roof, like this example that the author was lucky enough to sample, and the third was a convertible. The XJ41 project, which would have replaced the XJS, was cancelled in spring 1990.

One of the Collection Rouge special edition models produced for the American market during 1989. A special saloon, the Vanden Plas Majestic, was launched at the same time.

The Jaguar XJ-S Convertible has billions of extras as standard.

For the first time since the legendary E Type, Jaguar has brought back the convertible.

A Jaguar open to the sun, moon and stars.

The XJ-S V12 Convertible offers once again, the exhilarating rush of wind and sky above a snug cockpit. With the added comfort of knowing that the electrically operated hood can enclose you in just 12 seconds.

Performance is ample, to put it mildly.

The XJ-S Convertible and Coupé are powered by the road-going V12 from which Jaguar developed its 368 km/h racing engine.

Winner of the 1988 Le Mans 24 hour race, and two World Sports Car Championships.

For more information on the Jaguar range, please send your business card to Jaguar Australia, Freepost 15, PO Box 59, Liverpool, NSW 2170.

Or telephone Sydney 908 0822. Outside Sydney (STD free) 008 252 022.

LE MANS WINNER 1990

JAGUAR

Australian advertising dating from late 1990, and depicting the long-awaited XJ-S Convertible (seen here in V12 form).

Over the years, the XJ-S has attracted the attention of a massive number of companies wanting to modify the machine. This advert shows the Wide-Bodied Convertible produced by Paul Banham of England.

WIDE APPEAL

THE XJ-S WIDE BODIED CONVERTIBLE

A fully finished conversion by Paul Banham for any XJ-S coupé or convertible, with a choice of replacement aluminium or fibreglass front and rear wings.

Our range of superbly engineered Jaguar conversions also includes XJ-S Convertibles and Cabriolets, XJ-S round instrument conversions and XJ6 and XJ40 Cabriolets.

PAUL BANHAM CONVERSIONS
Unit 3, Charter Matrix
Business Park, Dartford, Kent
Tel: 0322 222969 Fax: 03225502

PAUL BANHAM CONVERSIONS

After an interim upgrade in 1991, the XJS was completely updated in May 1993. The Coupé now came with the four-litre, six-cylinder engine or the new six-litre V12, whereas the main change on the Convertible (which could also be bought with either power unit) was the 2+2 seating arrangement; previously, the drophead models had been strict two-seaters.

A 1996 model year XJS Convertible (this is a four-litre Celebration). The beautiful interior, which had changed considerably over the 21 years the S was built, can clearly be seen in this overhead shot, as can the 2+2 seating and the final rear light treatment.

The 1996 model year XJS four-litre Celebration models, seen here in both Coupé and Convertible forms. XJS production finally came to an end in April 1996.

Introduced in 1996, the XK8 was the latest in a long line of sporting Jaguars. Behind it, we can see examples of the undoubted pedigree of the marque, with the XJ-S, E-type, XK120 and SS100.

1

THE JAGUAR STORY

"I've never driven one and the comments I've heard up to now were almost uniformly 'beastly little car'. The late Malcolm Henderson, of Craig & Rose, the paint people, bought one for his honeymoon in, I think, 1934, and was so disgusted with it that he refused an offer of SS shares when the public company was floated. When I last spoke to him (1953, I think) he was still kicking himself for an error of judgement." - **Historian Michael Sedgwick talking about the SS2.**

To trace the history of Jaguar, one must go back to the days of the Swallow Sidecar Company, co-founded by William Lyons in the northern seaside resort of Blackpool. Born in September 1901, Lyons had always had a passion for motorcycles and cars, and worked for Crossley Motors and a local Rover dealership before seizing on the opportunity to go into business for himself with a friend called William Walmsley.

Walmsley had for some time been building sidecar combinations based on ex-War Department Triumphs in his parents' back garden, but only in a very small way. It was Lyons who had the nose for business, and, after many late night discussions, the pair eventually became partners, agreeing to acquire suitable premises and market the sidecars properly.

The Swallow Sidecar Company was formed in the early part of 1922. The demand for Swallow sidecars was such that extra production facilities were needed within a very short space of time. Thomas Walmsley, William's fa-

Sir William Lyons (1901-1985) - creator of the Jaguar marque.

Swallow sidecar production during the late 1920s.

An Austin-Swallow production model. The prototype had cycle wings, and the earliest cars had a slightly different windscreen to body fitting.

ther, duly bought a large building in Cocker Street and leased it to the pair. At the same time, the name of the business was changed to the Swallow Sidecar & Coachbuilding Company, and a new trim and paint shop was installed to justify the new title.

Already Swallow was becoming well-known for its prowess in competition, and a number of famous bike manufacturers were more than willing to display Swallow sidecars alongside their own machinery at the major shows. Among these was the Brough concern; George Brough later became an avid Jaguar fan.

Cyril Holland, who had served his apprenticeship at Lanchester, moved to Swallow in late-1926, bringing with him a great deal of coachbuilding experience. It was Holland who sketched what was to become the first Austin-Swallow, and built it on a Seven chassis supplied by Parkers of Bolton. The first car was completed in the spring of 1927.

Swallow went on to build its own coachbuilt bodies for a number of other popular cars, such as Morris, Fiat, Alvis, Wolseley and Swift. These proved immensely popular, and when Henlys of London put in an order for 500 Austin-Swallows, it was decided that the business had to expand. With the centre of the British motor industry very much based in the Midlands, it made sense to Lyons for the company to move southward and establish new and larger premises close to suppliers.

The move to Coventry

An old, disused gun shell-filling factory on Coventry's Whitmore Park Estate was leased for the sum of £1200 per annum - although the building had not been used for around ten years at least the price was right! Two adjacent blocks were also made available should Swallow require them in three years' time. It was the perfect set-up and the complete Swallow operation moved to Coventry in November 1928. In no time at all, the bank and the two sets of parents were completely paid off.

By far the most important link Swallow had made with other manufacturers was that with the Standard Motor Company. The first Standard-Swallow made its debut in October 1929, and was subsequently modified early in 1930 to feature the SS-style radiator grille.

In July 1931, information was leaked to *Autocar* magazine that Swallow would soon be known as "SS" and that a new two-seater coupé would appear for the 1932 season. Panic ran through the factory but the car was ready in time for Olympia; furthermore, the SS1 was undoubtedly the star of the show. Based on the six-

The original SS1, introduced at the 1931 Olympia Motor Show. Sadly, the smaller Standard Little Nine-based SS2 that ran alongside this model was not as successful.

cylinder Standard Sixteen chassis, it was a long, low and undeniably pretty sports coupé; also, unlike the earlier Swallows, it had the performance to match its looks.

Sales went well and, in late October 1933, Lyons registered a subsidiary company - SS Cars Limited. However, after Walmsley resigned from Swallow in January 1935, Lyons made SS Cars into a public company with himself in sole control. The Swallow Coachbuilding Co. (1935) Limited was duly formed to deal with the sidecar business, and so it was that SS was officially made the dominant marque of the business.

Jaguar is born

Lyons had fulfilled an ambition, but was still not entirely happy with the fact that the Standard name was still mentioned in magazine articles and such like when really the company was nothing more to SS than a supplier. The name Jaguar was suggested to him and, in September 1935, the new SS-Jaguar range was launched.

The SS100 - perhaps the car that more than any other before it elevated Jaguar to legendary status - made its debut in March 1936, having been announced six months earlier. The SS100 was powered by a 2.7 or 3.5 litre straight-six, and the motoring press couldn't praise the vehicle highly enough. With this model, a stunning competition record and the discreet elegance of the SS-Jaguar saloons, Lyons' company had become very highly respected by the time that England went to war.

The 1938 model year 3.5 litre SS-Jaguar saloon. There were subtle changes to the wing line, radiator shell and waistline moulding; the outside spare wheel mounting disappeared.

During World War II, Jaguar helped towards the allied war effort by producing thousands of aircraft components, as well as carrying out service and repair work on finished aeroplanes. Numerous War Department vehicles were also built and, at last, the final links with Standard were severed when Jaguar bought all the machine tools for the production of the six-cylinder engines. By the time war ended, Jaguar Cars Limited was established, the SS name being dropped due to the less than savoury images it conjured up after the conflict.

Lyons and his team wasted no time in introducing new models after the armistice (announcement of the post-war range of Jaguars came on the 21 September 1945), but perhaps their most important feature was the power unit. Christened the XK, this new twin-cam engine first appeared in four-cylinder guise, but was soon modified to

One of the stars of the 1948 Earls Court Motor Show, the XK120 was originally produced as a roadster, with the fhc arriving in March 1951. This picture shows a dhc (introduced in 1953) for the home market, with an fhc and a number of Mk VIIs in view.

Originally a Daimler Shadow factory, Jaguar moved to Browns Lane in November 1952; eight years later, Jaguar had bought the entire Daimler business. This picture was taken during the mid-1970s.

Le Mans 1953, with the victorious Rolt/Hamilton C-type in the pits. This win more than made up for the disappointment of the previous year.

become the familiar six-cylinder unit that remained in service for almost four decades. Making its debut in the beautiful XK120 sports car, the combination would go down in history as one of the world's finest post-war motor cars.

More competition glory followed and, with the launch of the Mk VII saloon, Jaguar once again had to expand. With permission to build between the existing factory and Beake Avenue being refused, negotiations began between Jaguar and Daimler about

The D-type was a potent replacement for the C-type, acquiring an enviable reputation in the field of competition. It's seen here on its way to victory at Le Mans in 1957.

II came along in October 1959. By this time, William Lyons had been made Sir William and again Jaguar found itself desperately short of space. The Daimler Motor Company could once more provide the solution to the problem.

The Daimler takeover

Despite putting on a brave face, Daimler was, in fact, struggling, losing vast sums of money due to a highly-restricted product range. Contracts for the Ferret armoured car were dwindling, bus production was down to ridiculous levels, and the only saloon cars available were either too large or outdated for volume sales. There was still no DN250 saloon on the horizon, and the SP250 sports car had simply not lived up to sales expectations.

With Daimler thus out of favour with its owner, BSA, the old Coventry firm was easy prey for Jaguar. Midway through June 1960, Jaguar purchased the whole of the Daimler business, along with the Radford factory, from the Birmingham Small Arms Company for the sum of £3.4 million. This allowed Jaguar to increase overnight production facilities and skilled workforce, and the company not only inherited an established prestige name but some superb V8 engines to boot.

Jaguar was quick to point out its position with Daimler, stating: "Jaguar Cars Limited wish to deny unfounded rumours to the effect that sweeping changes, including even the extinction of the Daimler marque, are to be expected. Whilst one of the most obvious benefits accruing to Jaguar as a result of this purchase is the avail-

The short-lived Mk IX arrived in October 1958 and, although it looked very much like the Mk VIII that went before, it was powered by a 3.8 litre XK engine instead of the old 3.4 litre unit.

the possible lease of the old Number Two Shadow Factory occupied by the latter company. This was situated in Browns Lane, Allesley, and, apart from completion of a few military contracts, was hardly used. Eventually Jaguar obtained the lease, and moved its entire operations across town from Swallow Road to Browns Lane in a relatively short period of time. The move was completed in November 1952 and the

company's head office is still there today.

The Le Mans 24 Hour race had fallen to Jaguar in 1951, 1953, 1955, 1956 and 1957, forging the marque's image and giving the firm a unique place in British motoring history, but it was not dependent on competition cars. The new saloon, known nowadays as the Mk I, gave Jaguar a tremendous sales boost, bettered only when the Mk

Following on from the successful 'Mk I', the Mk II compact saloon became one of the best-loved of all William Lyons' creations. This picture shows Jack Sears leading Colin Chapman in the 1960 July GP meeting at Silverstone.

The sale of the Daimler marque started a spate of purchases by Jaguar. The main reason for the company taking an interest in its Coventry neighbour was to enable it to acquire the Radford factory seen here.

William Lyons built up an impressive team around him. From left to right we can see some of the key people behind Jaguar's success: Harry Mundy, Walter Hassan, William Heynes and Claude Baily, pictured here with the XK engine, the heart of all Jaguars for nearly forty years.

ability of much-needed additional production facilities for Jaguar cars, the company's long-term view envisages not merely the retention of the Daimler marque, but the expansion of its markets at home and overseas."

Further growth

In 1961, Jaguar bought Guy Motors in an attempt to break into the commercial vehicle market; two years later, Coventry Climax, manufacturer of such diverse products as fork-lift trucks, fire pumps and some of the finest engines in Formula One history, was added to the Jaguar camp. Sir William also had a fine management team behind him, with F.R.W. 'Lofty' England, Walter Hassan, Harry Mundy, Arthur Whittaker, Bob Knight and Bill Heynes, to name but a few. It was, in effect a 'dream team', put together over a great many years.

Ironically, this was part of Lyons' problem - Edward Huckvale, who had been with the company since 1929, had already died, and none of the board was getting any younger. Sir William was looking towards retirement, and all of his most trusted colleagues were now far too old to take the helm. They could not be expected to stay on with the company indefinitely, even if they wanted to.

Another key member of the Jaguar management team was Lofty England, seen here on the far left viewing the rebuilt XJ13, a stillborn project initiated to take on the might of Ford and Ferrari at Le Mans.

Briefly, Lyons looked at taking over Lotus Cars. Some people on the inside say it was not so much for the company's products, as to get Colin Chapman on the Jaguar board in order for him to take over when Sir William retired. Sadly, Chapman wanted no part of running a company like Jaguar - after all, he had little to do with running Lotus at the time, his interest being centred on the Formula One racing programme.

March 1961 saw the launch of a legend - the Jaguar E-type. This is the fixed-head coupé model; without doubt one of the most beautiful cars ever built.

The Mark Ten saloon, built between 1961 and 1966. It was followed by the slightly modified 420G, which stayed in production until mid-1970, and formed the basis for the forthcoming XJ6 and the DS420 limousine. The Mark Ten was not only massive, but also a very advanced piece of engineering, setting the tone for future Jaguar saloons.

Lyons had to decide what was best for Jaguar.

British Motor Holdings

Sir William Lyons honestly believed that there was strength in a united British motor industry, so felt that a merger for Jaguar Cars Limited was in the best long-term interests of the company. With Jaguar so strong, he was in an ideal bargaining position. The first step was to find a suitable partner.

The British Motor Corporation had been formed in March 1952, and was basically a merger between Austin and Morris. As Brian Whittle put it in *The Morris Story*: "It signified the coming together of two of the industry's most bitter rivals. Why the two companies never joined before can be put down to a simple clash of strong personalities, not commercial reasons." However, BMC had done well through shared technology and facilities. Lyons saw BMC as an ideal candidate, and in December 1966, Jaguar merged with BMC to form British Motor Holdings.

This was now a huge concern, but for the first time since its formation, the BMC side of the business made a very poor return for 1967. Weakened by this, BMH was easy prey for the even bigger Leyland Group, which by now included Rover and Triumph. BMC had, in fact, held talks with Leyland earlier regarding a possible merger (before the deal with Jaguar had been struck), but now it looked like an easy takeover was on the cards.

British Leyland

In January 1968, Sir William Lyons gave up his position of Managing Director of the Jaguar Group, but remained Chairman and Chief

The Daimler DS420 limousine was the first product to come from the British Leyland Motor Corporation.

Executive. In the same month, the news of a BMH and Leyland merger broke. In his excellent book, *Jaguar - The history of a great British car* (Patrick Stephens Ltd), the late Andrew Whyte noted: "It was not what Sir William Lyons had wanted. He had hoped that the merger between Jaguar and BMC would provide the ideal answer for his company ..."

Putting on a brave face, a statement was released which said "Both BMH and Leyland are convinced that the merger is in the best interests of the country, the companies and all their employees." It was also noted that Jaguar would maintain its autonomy; the one thing it didn't say was for how long!

The British Leyland Motor Corporation began business in May 1968, and the first vehicle to come from this new concern was, in fact, a Daimler - the DS420 limousine. Designed to replace the old Austin Princess and the Daimler Majestic-Major limousine, it was announced on June 11. Built on a lengthened Jaguar Mark Ten (or 420G) floorpan and powered by the 4.2 litre XK engine, the bodies were finished at the Vanden Plas works in Kingsbury, Vanden Plas having been taken over by Austin in June 1946.

However, the DS420 pales into insignificance when compared to the next Jaguar-Daimler model. For years, Jaguar had followed a policy of producing many different models, but the latest car - the XJ6 - would eventually replace the entire range. Launched on September 26, 1968, the planning was again far enough advanced for it not to be interfered with by the Leyland heirarchy. The launch at the Royal Lancaster Hotel in London was as big a success as the car was, and the new model deservedly acquired many accolades.

In the meantime, Leyland boss Donald Stokes (by then Lord Stokes) was beginning to determine policy more and more, until eventually, just one month after its fiftieth anniversary, Jaguar Cars Limited failed to exist as a separate company. Stokes could see no point in producing Daimler V8 engines, as they were labour intensive to make and therefore too costly. The fact that the 2.5 and 4.5 litre units were good in service meant little, and it's even more galling when one thinks of all the support that the Mini received - for years and years, every one that was produced lost money for the company because no-one had bothered to check just how much it cost to build.

The XJ saloon was launched in September 1968. This picture shows the V12 version which made its debut in July 1972, and was instantly declared the fastest production saloon in the world: there was also a Daimler model known as the Double-Six.

As the situation worsened and the Leyland grip grew tighter, many people took the opportunity to leave. Bill Heynes had retired in mid-1969, and both Wally Hassan and Bob Grice left in May 1972. Lofty England had become Deputy Chairman following Arthur Whittaker's retirement, and became the Chairman of Jaguar in March 1972 after Sir William retired.

Lofty England was one of the few left with a link to Sir William's days, and was a natural successor to Lyons. Unfortunately, he had not been Chairman long when the workforce went out on a ten week strike over pay - nothing unusual, one might say, for the British Leyland days, but until now, labour relations at Jaguar had been surprisingly good. Regardless, the new XJ12 (the XJ6 with the V12 engine in it - first seen in the Series 3 E-type of 1971) received the 'Car of the Year' award, and England named the Daimler version the Double-Six after the elegant pre-war models.

With retirement looming, England had been looking for someone to replace him as Chairman, but had no luck. Anyway, it looked now as if he would have no choice in the matter, as Geoffrey Robinson was made the new MD of Jaguar by Stokes. Robinson arrived in early September 1973, just in time for the Series 2 XJ range to be launched at the Frankfurt Show.

Born in 1939, and therefore only 34 years old at the time, Robinson had risen through the Leyland ranks at an astonishing pace, acting as BL's financial controller before taking responsibility for Innocenti in Milan - part of the Leyland empire since May 1972.

In an interview with the *Coventry Evening Telegraph*, he said: "As Chief Executive, all executive responsibility is invested in me. Lofty and I have an excellent working relationship and there are many areas where his advice and experience are necessary." In reality, Geoffrey Robinson was very much his own man. Bill Heynes once said of Robinson: "He had remarkable drive, and was full of ideas and enthusiasm. We thought he might turn out to be another William Lyons or Billy Rootes."

Seeing that Robinson had a different approach to business, Lofty England retired early in January 1974. In all fairness, Robinson did do his best for Jaguar, trying to get a massive investment plan passed and creating a 'Management Board' in an attempt to get the Coventry firm "geared up for modernisation and expansion", but a major upheaval in company policy was about to take place.

The Ryder Report

By October 1974, with a record year for production having just been posted, it looked as if Robinson would get his

wish to expand the Jaguar operation with £12 million put by for a new body plant (body production was one of the main things letting down Jaguar quality at the time). Then Sir Don Ryder was commissioned by the government to do a report on Leyland's future, thus putting the whole idea on hold.

Briefly, the Ryder Report stated that all the companies within the British Leyland empire should be run from a central office. There would be one policy for all manufacturers of Leyland cars to follow, and so only local management was needed at each of the plants - all the major decisions would be taken by people miles away from Coventry who had little or no idea of what made a company like Jaguar succeed. Dealerships would become BL distributors and, overnight, many of the entrepreneurial dealers who could not afford to stock the whole range lost their franchises.

Jaguar would soon lose its sales and marketing freedom, with policy being dictated by BL, and there was no place for a Chief Executive of a company that, under the Ryder recommendations, no longer existed. After

the Ryder Report was published in the spring of 1975, Robinson, not getting the funding he needed to make Jaguar work, resigned and went into politics. An 'Operating Committee' was formed, led by Derek Whittaker and Tony Thompson, a Cowley man. And where did all the investment money go? Into developing the Metro ...

It was a crazy situation. One senior executive at Leyland International said: "Let me tell you. From a sales and service point of view, there's no difference between a Jaguar and a Mini - they're just bloody motor cars." At the same time, a Jaguar exhibition held at Coventry's Herbert Art Gallery to celebrate the company's Golden Jubilee year brought in more people than any other event held there. With the greatest of respect to owners, it is doubtful whether an exhibition of the Morris Marina or 1100 could have drawn the same crowds! Surely anyone could see that Jaguar was special, and deserved to be treated so?

Keith Cambage, who received his gold watch in 1976 with "Leyland Cars" inscribed on the back, said "My thoughts on the Ryder Report are un-

printable," which probably says it all. A dyed-in-the-wool Jaguar man who fought behind the scenes to keep the marque alive, Keith was part of John Egan's 'Task Force' and later went on to save the Daimler limousine, extending its life year after year, having brought quality up to standards fit for Royalty once again.

Maurice Ford, on the service side of the Jaguar-Daimler business, described the Leyland period as a "complete and utter disaster - too much politics and not enough management. There were times we were so unsure about what was happening and who we were working for, there was nothing we could do ... There seemed to be a new five year plan every six months."

The problem was further exacerbated by the in-fighting caused by years of marque loyalty. Cowley men thought little of Jaguar and the Rover-Triumph brigade, so tried to convert working practices to their own while the Jaguar people tried to protect them. John Barber, BL's Finance Director, had suggested the companies be kept separate, but no-one would listen. As quality dropped and the pound grew stronger against the dollar, export figures reflected a gloomy picture.

Passing of the E-type went by almost unnoticed, with the last Open Top Sports leaving the factory in February 1975 (the last 2+2 had been completed 18 months earlier). With American regulations dictating the end of the convertible, no direct replacement was announced, and the next car was a completely new Grand Tourer.

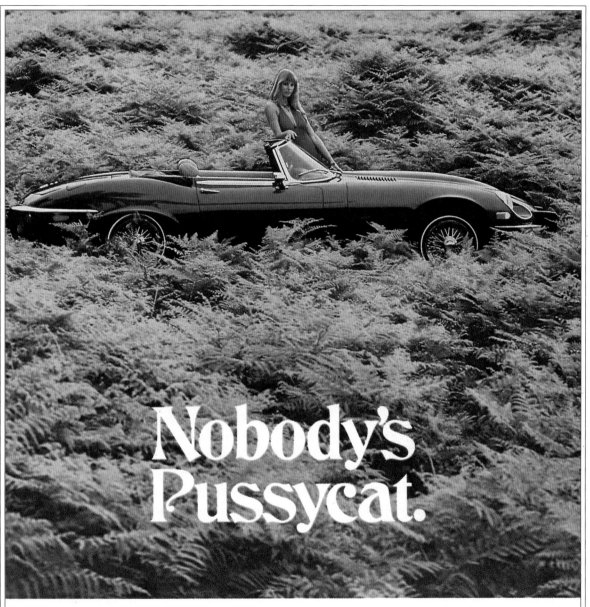

Nobody's Pussycat.

Of all the sports cars available to you, this is the one—the ultimate cat.

Because it offers what the others can't offer: the Jaguar V-12 engine.

And that changes the discussion from *what* a sports car can do to *how well* it can do it.

That's what the Jaguar E-type V-12 is all about. How well it glides from zero to fifty. How well it accelerates out of a pack and into the clear. Even how well it behaves in downtown traffic at quitting time before a holiday weekend.

In a word, the Jaguar V-12 is smooth. It's smooth going up the scale from zero and it's smooth going from cruising speed to passing speed. It's even smooth waiting for the light to change.

Because, from an engineering viewpoint, the Jaguar V-12 is in perfect balance. Since its 5.3 litres of capacity are divided by twelve—not eight or six—the forces are spread more evenly over the crankshaft by delivering smaller but more frequent pulses of power.

What is the effect like? Well, it's something like a turbine. And it's something like an express elevator. But it's not *exactly* like anything else. That's why you have to drive a Jaguar E-type V-12 before you decide on anybody else's sports car.

Since it is a Jaguar, it has independent front and rear suspension with "anti-dive" control. Power-assisted rack and pinion steering. Power-assisted disc brakes on all four wheels—ventilated in the front. A four-speed manual is standard, an automatic is optional.

So see the Jaguar E-type V-12. It's the only production V-12 sports car in town. And that makes it second to none.

For your dealer's name and for information about overseas delivery, call (800) 447-4700. In Illinois, call (800) 322-4400. Calls are toll free.

BRITISH LEYLAND MOTORS INC., LEONIA, N. J. 07605

27

2

BIRTH OF THE S

"The XJ-S takes British Leyland Cars into entirely new and extremely lucrative markets. It is aimed at the most discerning motorists in the world who want the very best. On all counts, the XJ-S meets these requirements with sheer style that is unmistakably Jaguar." - **Derek Whittaker, Managing Director, British Leyland Cars.**

Ever since the end of the Second World War, America has consistently provided Jaguar with its biggest market. With this in mind, it's not surprising that the replacement for the E-type had to be suited first and foremost to American needs. However, times were changing in the States: as well as being determined to reduce exhaust emission levels, especially in California, people like Ralph Nader and his group had introduced safety concerns to the big picture.

In retrospect, some rules issued during the late 1960s and early 1970s were plain stupid, but at least safety awareness was increased, which can never be a bad thing. Apart from side-impact and some fairly ludicrous bumper regulations, which some manufacturers interpreted better than others (fortunately for Jaguar, the Series 2 XJ saloon was one of the more successful designs to come out in a period blighted by ugly pieces of add-on rubber bumper extensions), perhaps the biggest concern was over the future of the open car. Federal crash regulations were now extremely strict, leading manufacturers to believe that the soft-top's days were numbered. These fears influenced car design to such an extent that it was well over a decade before a totally new, high-volume convertible was launched.

With E-type production falling from around 170 to nearer 90 units a week, it seemed pointless trying to modify the existing vehicle to meet the changing regulations - with falling sales, the chances of getting a good return on the investment needed to develop a Series 4 model were slim to say the least. Besides, most people associated the E-type name with convertibles, and as the soft-top's future was uncertain in the American market, this rather limited what could be

A mock-up of what would have been the Series 4 E-type. Fortunately, perhaps, for a number of reasons, the idea was not developed.

The elegant coupé version of the XJ, this being a Daimler Double-Six two-door. Although the car was announced in September 1973, production was halted until mid-1975.

done with the aging design. Although a mock-up of the Series 4 E was produced, in reality, a totally new car was the only route left open to the Coventry company.

Naturally, Jaguar already had an excellent saloon line-up, and was scheduled to announce a two-door version at the time of the Series 2 XJ launch in September 1973 (even a prototype Series 1 model exists). However, sealing problems with the frameless side windows would delay production until April 1975, but in any case, the XJ27 project was going to spawn a very different beast, intended for a different sector of the market.

Malcolm Sayer, the ex-Bristol aerodynamicist who provided the Jag-

uar marque with so many memorable body shapes, carried out a few styling exercises based on the mid-engined layout, but the concept was never really considered for production. Instead, he pursued the idea of a traditional front-engined, rear-wheel drive Grand Tourer - the new coupé would eventually be christened the XJ-S.

Styling

Under the heading 'A developing dialogue between stylist and scientist', the design stages were described in a Jaguar press release as follows: "The development of the XJ-S is the story of collaboration between two men, each with a highly individual approach to motor car design - Sir William Lyons,

the instinctive stylist, and Malcolm Sayer, the scientist and disciple of aerodynamics.

"Throughout the development programme the accent was on aerodynamics - but not simply to create a sleek car that slipped easily through the air. Malcolm Sayer's work on the famous C-type and D-type sports racing Jaguars, and later the E-type, achieved this with outstanding results. He wanted to take the process further: to harness the aerodynamic forces so that they positively contributed to road-holding and stability.

"Wind tunnel tests on full-size cars at the Motor Industry Research Association (MIRA) laboratories showed that the airflow pattern over the body shape was very good. When a spoiler was added to the front end and an undershield fitted below the engine, there were dramatic reductions - of about 50% - in the tendency of the car

One of the mid-engined design studies carried out by Malcolm Sayer.

Earlier design exercises held a number of styling clues. The front end of this early XJ6 prototype bears a remarkable resemblance to the XJ-S.

The same styling buck in profile. Note the bonnet line and front door.

A poor quality but extremely rare shot of the XJ27 in its earliest stages.

A more advanced prototype, with virtually all of the details finalized.

to lift, and 10% less drag was recorded.

"The spoiler also reduced aerodynamic side forces and moved the centre of pressure back - a very desirable characteristic for stability.

"While Malcolm Sayer used the slide rule and wind tunnel as his principal design aids, Sir William Lyons drew on the experience of more than 40 years of creating Jaguar cars - from the SS1 of 1931 - and imprinted the unmistakable stamp of his own strong visual sense.

"For both men the XJ-S is the last in a distinguished line of Jaguar cars. Malcolm Sayer died in 1970, after 20 years with the company, having firmly set the style of the XJ-S for the continuing development programme. Sir William is now retired.

"He believes the collaboration between stylist and scientist worked well. 'We decided from the very first that aerodynamics were the prime concern, and I exerted my influence in a consultative capacity with Malcolm Sayer.

Occasionally I saw a feature that I did not agree with and we would discuss it,' says Sir William.

"'I took my influence as far as I could without interfering with his basic aerodynamic requirements and he and I worked on the first styling models together. We originally considered a lower bonnet line, but the international regulations on crush control and lighting made us change and we started afresh. Like all Jaguars, we designed it to challenge any other of its type in the world - at whatever price - and still come out on top.'"

In fact, there was a third person involved with the XJ-S design who wasn't mentioned in the press release but who should be acknowledged - Jaguar's then Chief Stylist, Derek Thorpe. In reality, by the time Thorpe and his team came on the scene, the XJ27 body was virtually in its final form.

Initial work on the E-type's replacement began as the 1960s drew to a close. Sayer's first drawings were sent through to Sir William in September 1968, but with the expectation of the convertible being banned in the States, only a closed coupé design - XJ27 - was developed further; like the fixed-head version of the Series 3 E-

The XJ-S bodyshell was based on a shortened XJ6C/XJ12C floorpan.

type, a 2+2 seating arrangement would be adopted.

In an internal memo, Sayer stated: "The image sought is of a low, wide, high-speed car, at least as eye-catching as those the Italians will produce, even if it means sacrificing some of the more sensible values, such as luggage and passenger space, silence, ease of entry, etc."

Thorpe suggested changing certain aspects of the design, and a clay was duly produced with Sayer's design on one side and stylist Oliver Winterbottom's on the other. Lyons gave his seal of approval to the Sayer proposal, and the XJ-S was born. George Thompson (who was later heavily involved with XJ40) and his colleagues in the Styling Department were still not happy, and continued to submit sketches, but to no avail.

It must be said Thorpe was never overly happy with some of the features he inherited (the so-called 'flying buttresses' for instance), but nonetheless, he kept the styling as true to the original as possible, with just a few detail changes where necessary.

Almost from day one, the XJ-S shape could be clearly seen in Sayer's sketches. Although the stylists at Browns Lane submitted numerous other designs, with softer, more Jaguar-like lines (some of which were very beautiful, resembling Maseratis and Ferraris of the period), Lyons continued to go with Sayer's design.

Sadly, XJ28, the soft-top version, was shelved. Although the convertible ban never actually happened in the end (the proposal was squashed in

Federal Court in 1973), the news came far too late for cars like the XJ-S and Triumph TR7; ironically, when drophead versions were eventually sanctioned, a lot of extra work had to be undertaken to strengthen the bodies, which were designed first and foremost as closed coupés.

The floorpan was borrowed from the XJ6, which straight away ruled out any thoughts of a lightweight, compact sports car. The standard wheelbase on an XJ saloon was 108.75 inches (2760mm). The two-door models shared this dimension, although the four-door models soon adopted a 112.75 inch (2865mm) wheelbase across the range - what used to be classed as the lwb specification. The XJ-S, however, was built on a shortened 102 inch (2590mm) wheelbase - the same as that used on the XK models, but it was still six inches (150mm) more than the original E-type, and the track dimensions were, of course, far greater due to the XJ influence.

Basically, the XJ-S employed an XJ6C/XJ12C floorpan with the rear bulkhead and suspension mounting points brought forward by 6.75 inches (170mm), and there was substantial strengthening in the front bulkhead and engine bay, helping to make the

shell far stiffer than the XJ saloon. To meet the 1976 Federal regulations on rear impact, the fuel tank was positioned above the rear suspension, side-impact bars were enclosed in the doors, and special bumpers were fitted to overcome the US ruling on low-speed collisions to cars for all markets (XJ saloons destined for the States were fitted with different bumpers to those for Europe).

Despite having the same frontal area as the XJ saloon, the XJ-S was surprisingly slippery through the air, recording a Cd figure of 0.39 in the MIRA wind tunnel. To put this in perspective, the contemporary XJ6 had a Cd of 0.48, whilst the 2+2 E-type wasn't that much better at 0.455.

Although some have said that the model's styling was controversial, the author firmly believes the XJ-S was a much better-looking car than the 2+2 E-type. Yes, the two-seater fixed-head models were truly beautiful but, rightly or wrongly, it is my belief that the earlier 2+2s always looked a little awkward, and the Series 3 only ever looked happy as an open car. As for the 'flying buttresses'; firstly, Sayer was determined to incorporate them for aerodynamic purposes; secondly, from many angles, they are quite attractive - they

The XJ saloon was heading in a different direction altogether. As well as Jaguar's in-house styling department, both ItalDesign and Bertone were asked to submit ideas for the XJ40 saloon. Work started in 1973, but the project was put on hold a number of times. What is interesting is the complete contrast between this 1974 ItalDesign proposal and the XJ-S.

were just unusual for a British manufacturer, but I've never heard any complaints about the styling of the Ferrari Dino - and thirdly, they give the car character: unlike so many modern automobiles, an XJ-S stands out from the crowd.

The bodyshell, produced at the Pressed Steel Fisher plant in Castle Bromwich, weighed in at 327kg (about 14% lighter than that of a lwb XJ saloon), but masses of sound-deadening material ensured the car was always going to be a heavyweight cruiser. Despite the sporting nature of the vehicle, noise suppression was one of the key aspects Jaguar's engineers were striving for - even the engine bay was tapered towards the rear to deflect noise from the power unit.

Inside, the lack of wood on the dashboard and door cappings, to be perfectly honest, spoilt an otherwise well-planned cockpit. This carried on a tradition started in 1957, where sport-

The magnificent V12 as it first appeared in the XJ-S. Engines (and manual transmissions) were built at the old Daimler factory in Radford, Coventry.

ing models - at that time, the XK150 - lost the walnut veneer fascia, but something was definitely needed to lift the ambiance in the XJ-S. If the trim colour was black, the interior was almost austere, despite the well-shaped leather-trimmed seats and high-quality carpets. Instrumentation was largely from the XJ saloon, although the gauges were unique to the XJ-S and quite innovative.

The engine

As mentioned earlier, the V12 made its debut in the Series 3 E-type in spring 1971 in a bid to restore power progressively being lost through ever-tighter Federal emissions laws. In all, over 72,000 E-types were built during a 14 year production run - 15,290 of them were Series 3 cars powered by the V12 engine.

Jaguar's V12 engine was first thought of in the early 1950s, but wasn't developed until the early 1960s when it was envisaged as the cornerstone of the company's plan to return to motor racing, which it had dominated - courtesy of the legendary C- and D-type - in the previous decade. The V12 project was handled by a team of engineers which included Bill Heynes, Claude Baily, Walter Hassan and Harry Mundy.

The first fuel-injected, 4991cc four-cam unit, which developed a healthy 502bhp on the testbed, was installed in a mid-engined monocoque racer, the XJ13. Tests in 1967 on the banked track at MIRA instantly produced a new 161mph lap record, with 176mph being attained down the straight.

Eventually, the plan to re-enter racing was scotched, but the V12 continued to be developed for road use. Three Mk X saloons had been fitted with the four-cam unit for testing in autumn 1964 (initial, rather ambitious, hopes had been for Jaguar to return to Le Mans in 1965), but for road car use, it was decided to employ a sohc per bank arrangement.

The two-cam, flat-head design was not only quieter and lighter, but was also found to perform just as well as the four-cam engine up to 5000rpm; it also produced a higher torque output at lower revs, which was an important factor, given that most of the new models would leave the factory with automatic transmission. Naturally, it was also cheaper to machine and less complicated to build and service.

Even for road use, it had been hoped to use fuel-injection from the start; the old Lucas system gave good performance, but apart from the fact the unit would never have met the contemporary Federal regulations, the metering system was not sufficiently accurate. Brico came up with the perfect solution - electronically-controlled fuel-injection - but cancelled plans at the last minute. Faced with little alternative, four carburettors were employed instead.

In March 1971, with a capacity of 5343cc, the V12 unit developed 272bhp (or 250bhp in Federal specification), and redefined industry standards for power, refinement and flexibility. Whilst the 4.2 litre XK unit had produced similar outputs in the past, power was now quoted at 157bhp (nett) in the States; the same kind of performance from the XK unit was just not possible with US anti-smog equipment fitted.

By the following July, the V12 had found its way into the XJ saloon (the car it was originally designed for). The XJ12 became the fastest production four-seater in the world, and was immediately voted 'Car of the Year'. From a very early stage, it was decided the only power unit for the XJ-S would be the V12.

When the Series 2 XJ saloons were announced in late 1973, apart from the obvious styling differences, the V12-powered, two-door models were specified with fuel-injection. The intention was to give better fuel economy and improved emissions control: in the latter half of the 1970s, an *Autocar* test showed a massive 21% improvement in fuel consumption over the carburettor V12 saloon the magazine tested some time before.

At last, engine specification for the XJ-S had come together: it was to

A cutaway drawing of the new Jaguar, signed for the author by Norman Dewis, the company's Chief Test Driver for several decades.

be an all-alloy, 60 degree V12 of 5.3 litres with fuel-injection. By the time the car was ready for launch, power was quoted at 285bhp DIN at 5500rpm - more than enough to endow the new Grand Tourer with true GT performance.

Mechanics of the new car

Mated to the V12, there was a choice of a four-speed manual or three-speed automatic gearbox. The manual unit was a slightly modified version of that used in the Series 3 E-type. Although a five-speed gearbox would have been preferred, there wasn't a suitable overdrive unit available at that time (Bob Knight had proposed an electronically-controlled, two-speed axle to overcome the problem, but it never made it to the marketplace). The automatic option came in the form of Borg-Warner's Model 12 gearbox, and proved to be far and away the most

The front suspension of the XJ-S.

popular form of transmission. With either gearbox, power was transmitted via a one-piece propshaft to a Salisbury final-drive unit incorporating a Powr-Lok differential.

Jim Randle, an ex-Rover apprentice who had been with Jaguar since 1965, and eventually became the company's Director of Engineering, worked long and hard at providing the XJ-S with the superb ride quality that has made Jaguar famous throughout the

The independent rear suspension (irs) being built up, seen here upside down. Note the inboard rear brakes.

Testing at MIRA, with Norman Dewis at the wheel. In all, ten prototypes were produced, most of which were either scrapped or written off. A number ran on public roads in camouflage - the nose and rear-end were covered to keep the shape a secret until launch.

brakes, and an anti-roll bar borrowed from the latest version of the XJ12 (ie at 22mm (0.87in) rather than the previous 19mm (0.75in)).

At the back, the XJ-S featured the, by now familiar, independent rear suspension, first seen on the E-type. The design was based on that of the double-wishbone principle, though the top wishbone was actually a driveshaft, and the lower one evolved into a tubular forked arm which carried a large aluminium hub where the driveshaft was located. Trailing radius arms attached to the body floor and the lower arm kept the unit in check; like most of the components, these were rubber-mounted. Two shock absorbers enclosed by coil springs were employed on each side. As four spring and damper units were used, each could be made smaller, making the whole rear suspension a more compact unit.

Solid inboard disc brakes were used at the back, attached to the differential with the driveshafts running from them. Most of the unit was enclosed within a subframe, which acted as a mounting point for the differential, and again, there was no metal-to-metal contact, the subframe being rubber-mounted for noise insulation. This suspension, introduced in 1961, was fitted to all Jaguar and Daimler models for many years to come, although the XJ-S did feature a rear anti-roll bar, which was unique to the Grand Tourer.

The standard alloy wheels came from the recently-introduced XJ5.3; at 7.9kg there was a saving of 2.8kg per wheel when compared to standard steel wheels, providing a useful 26% reduc-

world. At the same time, he gave the car more sporting handling characteristics and sharper steering response, compared to its saloon counterparts.

The suspension was basically the same as that found on the XJ saloons, although spring rates were adjusted to suit the lighter GT model - around 90lb/ins at the front and 125lb/ins at the rear. Up front, there was the traditional Jaguar double-wishbone set-up (with the lower wishbone acting as a spring pan), a single coil spring and damper each side, ventilated disc

tion in unsprung weight. However, the lighting arrangements were commissioned purely for use on the new GT; American specification models had twin-headlights, but other markets were provided with special, oblong-shaped units manufactured by Cibie. Thus, as far as the car was concerned, the scene was set ...

Launch of a legend

Despite Jaguar having had a good year in 1974, massive BLMC losses as a group led to the nationalisation of the company, and in 1975, it was renamed British Leyland Ltd. The state of the British industry was dire, with the government having to support Chrysler UK as well (the vestige of the old Rootes Group) in an attempt to save jobs. At the other end of the scale, with rising fuel prices and a favourable exchange rate against the yen, Japanese economy car sales were booming, and eventually an import limit was imposed to try and protect home producers. Amid the political turmoil and upheaval within the BL organisation, the launch of a 150mph supercar could, perhaps, have been timed a little better.

The XJ-S was introduced to a select group of the world's motoring writers in May, when four cars were made available for road testing in Britain, and then via a press release dated 10 September 1975. Entitled 'The Finest Jaguar Yet', it stated: "The most exclusive and expensive Jaguar ever produced - an exciting 150 mile-an-hour coupé called simply the XJ-S - is unveiled today (10 Sept 1975) by British Leyland. Launched on the eve of the Frankfurt Motor Show, it is designed to capture sales from that small jet-set group of European cars which can be counted on the fingers of one hand. The XJ-S combines a new and distinctive aerodynamic body with unsurpassed levels of comfort, refinement and quietness and the silky smooth power of the Jaguar V12 5.3 litre engine with electronic fuel-injection.

Another superb shot of JVC 484N. (Courtesy Paul Skilleter)

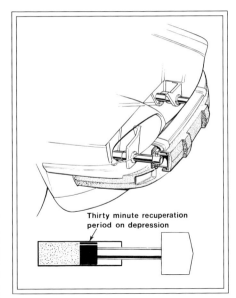

The Menasco struts fitted to the XJ-S bumper were one of the best designs produced during the 1970s. Meeting Federal regulations of the time often meant compromising a car's looks, but this was very neat.

"Although expensive for a Jaguar, the XJ-S will be modestly priced by comparison with its competitors - and will have the added advantage of easily-accessible service points around the world. The XJ-S is aimed at winning exports - particularly in the North American market - where it is expected to earn £26 million in the first year with other overseas sales adding another £4 million.

"Performance is outstanding with a 0-60 time of just 6.8 seconds, but the combination of a fuel-injected V12 5.3 litre engine with sleek styling produces excellent fuel economy for a car of this class. Development drivers returned between 15 and 18mpg.

"Production at the Coventry factory has initially been set at about 60 cars a week. The XJ-S incorporates several 'firsts'. New and powerful quartz-halogen biode headlights have been designed for the car by Cibie in conjunction with Jaguar. The tyres, too, have been developed exclusively for the XJ-S. The fuel, water temperature, oil pressure and battery condition gauges are of a new design for pinpoint accuracy, and there is an 'at-a-glance' warning system for instant driver reaction. The bumpers, the first of their type in Europe, are mounted on telescopic shock absorbers to prevent 5mph damage."

The technical details

The original press release was very informative by the standards of 1975, and after the usual introductory pages, proceeded to describe each aspect of the new Jaguar in minute detail. The contents have been reproduced here in full to ensure historical accuracy:

The XJ-S, the most advanced development of the XJ theme, incorporates many completely new features, several of which are not only new to Jaguar and British Leyland, but are a world first.

In this, the most expensive and exclusive production Jaguar, every effort has been made to produce the quietest, most comfortable and luxurious high performance car on the market. Particular attention has been paid to driver control layout, seating and interior silence.

Strength and safety

The XJ-S steel monocoque bodyshell meets all current safety regulations and anticipates forthcoming legislation in North America and Europe. All cars have the same bodyshell and exterior features, including 'no-damage' 5mph impact bumpers.

The bumpers are the first of this type on the European market, and

Interior of the XJ-S. The S-type was destined to become Jaguar's (and therefore British Leyland's) flagship model. As such, its role changed from sporting in the prototype stages to luxury GT by the time it was launched.

feature a steel armature mounted to Menasco struts housed in the longitudinal chassis members. Side intrusion barriers are built into the very wide doors, which are mounted on reinforced hinges.

The forward hinged bonnet gives excellent access to all major components and is supported by two gas struts. A dual-purpose lever both opens and positively locks the bonnet.

Luggage capacity in this car is exceptionally large even by luxury saloon car standards. Jaguar engineers have designed the rear end of the XJ-S not only to meet rear impact tests but also to accommodate as much luggage as the XJ saloons. The deep boot also houses the battery, fuel tank, fuel pump and the spare wheel, which has a 'tai-lor-made' cover for luggage protection.

Jaguar's meticulous attention to detail is illustrated by louvres in the boot which keep the air fresh and remove staleness.

Style with comfort

The interior of the XJ-S has been designed and developed with one main objective in mind - to give the owner the highest possible degree of sumptuous refinement and yet retain the taut feel of an exclusive high performance car.

New seats

Four adults can travel in comfort. The front seats are fully adjustable fore and aft, fully reclining, and of a new design with the seat cushion made up of two separate components, the centre section and the outer square. The centre section 'gives' under the weight of the occupant, while the outer part firmly grips without pinching.

The door and side panels have combined armrests, door pulls and individual ashtrays.

Inertia reel front seat belts are standard and are available for the rear seats, too. The reels are hidden behind trim panels. Rear seat belt buckles are neatly housed in a central tray.

The sound of silence

An established hallmark of the Jaguar XJ theme is interior silence, and the XJ-S can be firmly placed amongst the world's quietest cars.

The bulkhead and transmission

Rear seating was for occasional use. With the S being such a low car, entry and exit from the back wasn't easy, and when there, conditions were cramped for all but the smallest.

tunnel are fitted with close-fitting moulded heat and sound shields. Wiring passing through the bulkhead is by multi-pin plug and socket units [as on the Series 2 saloons] - to avoid holes through which noise can pass.

The basis of the interior insulation is the fitting of moulded undercarpet pads tailored to fit exactly into the front footwells, bulkheads and transmission tunnel. Damping pads are fitted to the floors, door panels and rear seat pan, and the parcel shelf and rear side panels are all sound insulated. The floor and lower door panels are trimmed in deep-pile, felt-backed carpet.

The entire luggage compartment is trimmed in black carpet, and a layer of sound insulation is sandwiched be-

tween the petrol tank and rear seat bulkhead and the axle cover.

Take notice - 18 times over

The full-width, one-piece fascia houses the instruments, multi-directional air vents and a lockable glove compartment. The fascia unit is vacuum-formed and trimmed in black PVC with a thickly padded crash roll along the top. A lower central console houses the auxiliary switches, clock, radio and the heating and ventilation controls.

The instruments are all in a single nacelle which contains speedometer, tachometer, fuel, water temperature, oil pressure and battery condition indicators, and 18 warning lights along the top. All major functions, both me-

chanical and safety, are monitored by the warning lights. The instrument binnacle is wired by printed circuits, and two multi-pin plugs connect it to the main wiring loom. The rim of the 15.5 inch steering wheel is covered in leather. The wheel is an acrylic moulding which resists chips and scratches and is lighter than normal steering wheels.

Warning - at a glance

For XJ-S drivers red and amber mean more than a wait at traffic lights - they are the key colours in an 'at-a-glance' system of warning lights to give the driver immediate warning of a major failure, a secondary fault and information about which auxiliary equipment is operating.

40

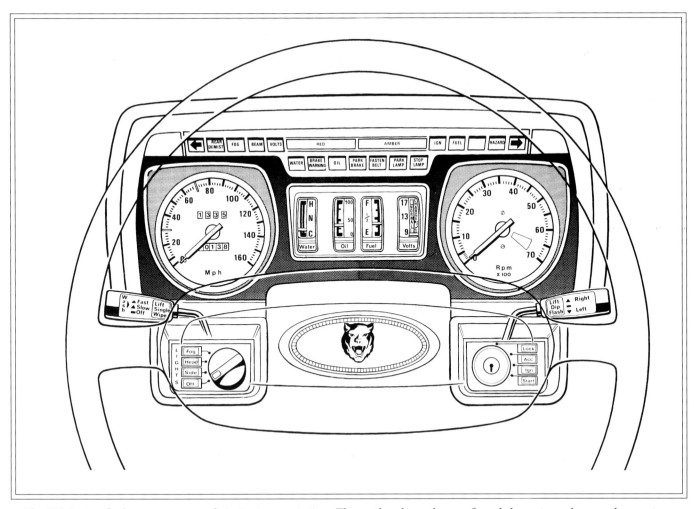

The XJ-S signified a new approach to instrumentation. The author has always found the set-up clear and easy to understand, although the turn lights can be hidden by the rim of the steering wheel.

Auxiliary push-push switches operate the heated rear window, interior lights, map lights and hazard warning lights. Ignition and lighting controls are located in the fascia on either side of the steering column and are illuminated by the same fibre-optic system that was introduced on the Series 2 XJ saloons.

The fascia incorporates a large lockable glove compartment with a flip-up vanity mirror in the lid. The sunvisors are recessed into the one-piece headlining.

Electrically-operated windows and central door locking are standard and the controls are on the gearbox console.

The XJ-S has five interior lights -

two behind the rear passengers, one at each end of the fascia and one above the driving mirror.

The handbrake lever is mounted on the driver's door sill. When the handbrake is applied the lever can be returned to the 'off' position, out of the way. The brake is then released by pulling up the lever and pressing the release button in the top of the handle.

Automatic temperature control

The XJ-S has the fully automatic air conditioning developed by Delanair and Jaguar for the Series 2 XJ saloons. There are only two controls, one for temperature and one for the automatic and manual operation.

The system provides a wide range

of temperatures and the occupants select the required setting which is then maintained regardless of outside conditions.

Ventilation is through fascia and footwell vents. The fascia has four multi-directional vents, one at each end of the fascia and one on each side of a large central vent. The footwells are ventilated from the sides of the transmission tunnel.

The air conditioning is programmed to maintain the face level air at a lower temperature than the footwells to ensure clear heads and prevent stuffiness. Air extraction is through the rear quarter panels via a collection chamber in the body panels. The chamber is lined with foam to

A left-hand drive XJ-S (but sporting British number plates) at speed. This was one of the original press pack photographs.

silence roar as the air leaves the car.

Power to match power

In a high speed car like the XJ-S it is vital to have powerful, far-seeing headlamps. A world first for Jaguar and Cibie are new quartz-halogen biode headlamps developed for the car. The lamps have two independent reflectors, each with a halogen H1 bulb, and the reflectors are specially designed for dip or main beams. To achieve the correct shape of the dipped beam cut-off and the widespread main beam, a screen of auxiliary prisms is located inside the main reflector. The screen, combined with the outer lens, provides a long, wide spread of light for fast, accurate driving.

Grace and pace

The heart of the XJ-S is the Jaguar 5.3 litre fuel-injected V12 engine, which produces 285bhp DIN and 294lbft DIN torque, and is renowned for its smoothness and flexibility. Lucas electronic fuel-injection has increased the power of the engine, but has given it a flatter torque curve to make it one of the most flexible and tractable power units in the world.

Capable of vivid acceleration, the XJ-S has a maximum speed of 150mph plus. Aerodynamics and fuel-injection combine to give fuel economy and Jaguar development drivers have returned a fuel consumption range of 15 to 18mpg.

Fuel system

The single 20-gallon tank is mounted in the safest position - between the passenger compartment and the boot, between the rear wheels. The tank feeds into a sump tank immediately below, designed to provide a continuous supply of petrol to the high pressure fuel pump, even if the main tank is running low. The fuel pump has a permanent magnet motor which runs immersed in petrol to the injectors.

Transmission

A choice of two gearboxes is available for the car: the Jaguar four-speed manual all-synchromesh unit or the Borg-Warner three-speed fully-automatic Model 12 transmission.

A single-piece propellor shaft drives the differential. This is a four-pin Powr-Lok unit with a 3.07:1 gear ratio.

The rough and the smooth

Sharp corners, bumpy lanes or a motorway - the XJ-S covers all in a manner no other car can match. Cushioned on the Jaguar all-independent suspension systems, ride

42

and handling are superb.

Very careful attention has been paid to the total roll stiffness and its distribution at the front and back. This has resulted in the fitting of new rear and uprated front anti-roll bars; at the same time a very critical look at the rear spring and damper units and the damper force/velocity relationship produced a rear suspension that has exceptional handling and a very comfortable ride.

A great deal of work was carried out on the steering geometry in conjunction with the new Dunlop SP Super Sport tyres. The result was the fitting of an eight-tooth pinion - giving a reduction in overall ratio [the saloons were fitted with a seven-tooth pinion]. This work produced extremely accurate steering with a very rapid response.

A new, fast tyre

The new tyres have been developed for wet road performance, quick response, low wear rate and to run at higher speeds than the SP Sport.

Jaguar and Dunlop returned to one of their old record-breaking haunts - the Jabbeke autoroute in Belgium - to carry out high speed testing of the tyres and the car.

Accompanying the new tyre is the latest light alloy road wheel from Kent Alloys, recently offered for the first time on XJ5.3 saloons.

The new wheel, designed exclusively for Jaguar cars, is a low pressure die-cast unit made from a development of BSS1490 LM25 aluminium alloy, further developed by GKN Kent Alloys to improve elongation and give an ultimate tensile strength of 14 tons per square inch.

Reaching a full stop

In just over 20 seconds, the XJ-S can accelerate from rest to 100mph and brake to a stop. To give this stopping power the front brakes are 11.18 inch diameter ventilated discs with four piston calipers and on the rear, mounted inboard, are 10.38 inch discs. Friction material is Ferodo 2430, which is renowned for its high coefficient of friction and long life with very good anti-fade characteristics.

Incorporated into the front to rear split hydraulic circuits is a pressure differential warning actuator, which causes the brake warning light to illuminate if one circuit fails.

Servo assistance is fitted in the form of a Girling in-line tandem servo mounted on the brake pedal box. This is direct-acting, using vacuum assistance from a vacuum reservoir tank located under the front wing.

Wipers

The two-speed windscreen wipers are controlled by a steering column stalk with a single wipe facility. The motor linkages are mounted as a single unit underneath the windscreen scuttle air vent, and the entire unit lifts out after removal of four screws. The simple mounting arrangement, for reliability and easy maintenance, is another 'European first' for the XJ-S.

Electronics and electrics

To equip a car like the XJ-S with electronic fuel-injection, electronic air conditioning, quartz-halogen biode headlamps, electric windows, central door locking, heated rear window, two-speed wipers, 18 warning lights and all the normal electrical equipment, needs a highly sophisticated wiring system. All components are wired by multi-pin plugs, which are coded in the layout of the plug pins.

Feeding power to all these systems requires a high capacity battery and a powerful alternator. Lucas provides both with the Pacemaker CP13 battery with a capacity of 68 amps at a 20 hour rate, and the 20 ACR alternator producing 60 amps at 3500 engine rpm.

Rear lights

The rear light clusters wrap around and combine rear, stop and flasher lights, reflectors and side markers. Reversing lights are in the number plate illumination panel which runs the full length of the bootlid.

Exhaust system

In common with the rest of the Jaguar range, the exhaust system has stainless steel silencer boxes, tail pipes and over-axle pipes.

Specifications

Engine
Type	Four-stroke petrol engine, water-cooled
No. of cylinders	12 in 60 degree Vee
Bore	90mm (3.54in)
Stroke	70mm (2.76in)
Capacity	5343cc (326 cu.in)
Piston area	763.2cm^2 (118in^2)
Compression ratio	9.0:1
Stroke/bore ratio	0.78:1

Performance
Power	285bhp DIN at 5500rpm
Torque	40.7mkg (294lbft at 3500rpm)

Cylinder block
Type	Open deck
Material	Aluminium alloy LM25
Pistons	Aluminium alloy solid skirt with combustion chamber in top
Piston rings	Three - two compression and one multi-rail oil control
Crankshaft	Three plane, seven main bearings. Tuftrided manganese molybdenum steel

Cylinder heads
Material	Aluminium alloy LM25 WP
Camshafts	Two - one per bank
Valve layout	Single overhead camshafts operating bucket type tappets
Valve lift	9.525mm (0.375ins)
Tappet type	Inverted bucket
Tappet clearance	Inlet 0.30 - 0.35mm (0.012 - 0.014in)
	Exhaust 0.30 - 0.35mm (0.012 - 0.014in)
Valve timing	Inlet 17 degrees BTDC, 59 degrees ABDC
	Exhaust 59 degrees BBDC, 17 degrees ATDC

Sump
Type	Steel pressing with internal baffles

Lubrication
Type	Pressure
Pump type	Internal and external gear with crescent type cut-off
Normal oil pressure	4.9kg/cm^2 (70psi)
Filter	Full-flow paper element

Ignition
Type	Lucas Opus Mk II electronic
Firing order	1a, 6b, 5a, 2b, 3a, 4b, 6a, 1b, 2a, 5b, 4a, 3b
Distributor	Lucas magnetic impulse type
Ignition timing	750rpm, 10 degrees BTDC
Sparkplugs	Champion N10Y
Gap	0.625mm (0.025in)

Injection
Type	Lucas electronic manifold injection
Enrichment	Automatic cold start injector
Induction manifolds	Two six-branch aluminium alloy

Fuel system
Type	Recirculating
Pump	Lucas electric permanent magnet motor
Fuel specification	97 octane - four star

Electrical equipment
Polarity	Negative
Battery	Lucas CP13
Battery capacity	68 amps at 20 hour rate
Starter	Pre-engaged
Alternator	Lucas 20ACR
Alternator capacity	60 amps at 3500 engine rpm
Horns	Twin Lucas self-earthing

Cooling system
Type	Water pressurised. Impeller pump belt driven off crankshaft
Pressure	1.056kg/cm^2 (15psi)
Radiator	Marston Superpak crossflow
Thermostat	Two wax type opening at 82 degrees C
Fans	12 bladed steel fan with viscous coupling and thermostatically-controlled, electric four-bladed fan

Exhaust
Layout	Four downpipes merge into two double-skinned pipes. Two main and two rear silencers
Exhaust emissions (North America)	Exhaust emission controls incorporate exhaust port air injection, exhaust gas recirculation and a catalytic reactor for each bank
Evaporative loss	Engine anti-run-on valve. Vapour from the fuel tank is piped via a separator canister to a charcoal canister which is purged by manifold depression

Transmission
Manual gearbox	Four-speed all-synchromesh		
Clutch	Single dry plate		
Plate diameter	380mm (10.5in)		
Automatic gearbox	Borg-Warner Model 12 three-speed		
Torque converter	2.03:1 ratio		
Ratios		*Manual*	*Auto.*
	1st	3.238:1	2.39:1
	2nd	1.905:1	1.45:1
	3rd	1.389:1	1.0:1
	4th	1.0:1	-
	Rev	3.428:1	2.09:1
Axle ratios		3.07:1	3.07:1
Overall ratios	1st	9.94:1	7.34:1/14.68:1
	2nd	5.85:1	4.46:1/8.92:1
	3rd	4.26:1	3.07:1/6.14:1
	4th	3.07:1	-
	Rev	10.51:1	6.41:1/12.82:1

Brakes
Type	Disc brakes all-round
Layout	Dual circuit split front to rear with pressure differential warning actuator
Servo	In-line tandem vacuum servo
Discs - front	Ventilated cast iron, 284mm (11.18in) diameter
Discs - rear	Cast iron with damper ring in periphery, 263mm (10.38in) diameter
Calipers - front	Four-piston caliper
Calipers - rear	Two-piston caliper
Friction materials	Ferodo 2430
Rubbed area	Front: 1624cm^2 (252in^2)
	Rear: 956cm^2 (148in^2)
	Total: 2580cm^2 (400in^2)

Suspension
Front layout	Fully-independent, semi-trailing wishbones and coil springs. Anti-dive geometry. Girling Monotube dampers. Anti-roll bar
King pin inclination	1.5 degrees (+\- 0.25)
Castor angle	3.5 degrees (+\- 0.25)
Camber angle	0.5 degrees positive (+\- 0.25)
Alignment	0.312cm (0.063-0.125in)
Springs	Free length: 305-356mm (12-14in)
	Rate: 423lb/in
Anti-roll bar diameter	22mm (0.875in)

Rear layout	Lower transverse wishbones with driveshafts acting as upper links. Radius arms. Twin coil spring and damper units. Girling Monotube dampers. Anti-roll bar
Camber angle	0.75 degrees negative (+\- 0.25)
Anti-roll bar diameter	14mm (0.562in)

Wheels and tyres

Wheels	GKN Kent light alloy wheels to Jaguar design
Size	6JK rim, 38.1cm (15in) diameter
Tyres	Dunlop SP Super steel braced with block tread pattern
Size	205/70 VR15

Steering

Type	Adwest power-assisted rack and pinion with energy absorbing column
Wheel diameter	393.7mm (15.5in)
Turns lock-to-lock	Three
Overall ratio	16:1 with an eight-tooth pinion
Pump	Saginaw rotary vane

Body

Manufacture	British Leyland, Castle Bromwich
Type	All-steel monocoque construction. Two-door four seater with forward opening bonnet and large boot
Exterior features	Complete underbody protection. Driver's door mirror. Front air dam and undershield. Radio aerial. Flush-fitting filler cap cover. Recessed door handles
Bumpers	Wrap-around front and rear bumpers consisting of a steel armature mounted to Menasco struts and with a synthetic rubber cover. Designed to meet 5mph impact tests
Glass	Laminated windscreen with toughened side and rear windows. Sundym glass is standard. Electrically-heated rear window
Locks and keys	Doors are fitted with high anti-burst load locks with flush fitting interior and exterior handles. Electrically-operated central door locks. Separate keys for ignition, doors and boot and glove locker. Bonnet release and locking lever below fascia

Interior

Seating	Two fully-reclining front bucket seats with separate squab centre section. Head restraints. Two individual rear bucket seats
Upholstery	Seat facings trimmed in leather. Sides, backs and door trims in expanded PVC. Nylon pile carpets for total floor area and lower door panels. Interior is trimmed throughout in flame retardant materials
Seat belts	Inertia reel lap and diagonal. One-handed operation locking buckles plug into quick release sockets
Interior equipment	Five push-button radio and door speakers. Dipping rear view mirror with break-away support. Padded sunvisors recessed into roof panel. Black leather grained fascia and

console. Combined door pulls and armrests. Individual ashtrays front and rear. Three interior and two map lights with switches and operated by door switches. Clock

Electrical equipment

Instruments Instruments and warning lights set in fascia panel in a compact self-contained unit in front of the driver. Auxiliary instruments are 'air-cored' indicators
Speedometer calibrated 0-160mph with total and trip recorders
Tachometer 0-7500rpm (red line 6500rpm)
Fuel level indicator
Water temperature indicator
Oil pressure indicator
Battery condition indicator
Instruments are illuminated by diffused green lighting

Warning lights system Dual system of warning lights with two 'attention getters' giving warning to stop immediately (red) or investigate when convenient (amber):
Indicators (two)
Rear demist - heated rear window on
Main beam - main headlamps on
Fog -fog lights on
Volts - indicates battery overcharge
Major fault warning - stop car and investigate
Secondary fault warning - investigate as soon as possible
Ignition - ignition on
Fuel - low fuel level
Hazard - all indicators flashing
Water - low coolant (red 'attention getter')
Brake failure - low fluid or failure (red 'attention getter')
Oil - low oil pressure (red 'attention getter')
Park - handbrake on (amber 'attention getter')
Fasten belt - seat belts not fastened (amber 'attention getter')
Park lamp - failure of any sidelight (amber 'attention getter')
Stop lamp - failure of brake light bulb (amber 'attention getter')

Controls and switches Fascia mounted master switches on either side of the steering column for lighting and ignition. LH column control for two-speed wipers, wash and single wipe. RH column control for headlamp flash, dip and full beam. Padded steering wheel boss acts as horn push. Push-push switches for hazard, heated rear window, interior and map lights. Two air conditioning controls in lower console for automatic control and temperature selection. Two rocker switches for electric windows and one for central door locking

Lighting	Headlamps: two Cibie biodes with two quartz-halogen filaments - all four operating on main beam. (North American cars have a four headlamp system with tungsten filaments)
	Wattage: 55/55 watts
	Side lamps: Combined side and indicator lamps housed in front bumper
	Rear lamps: rear light cluster in rear quarter panels with wrap-around lenses. Stop, tail, indicator and reflectors. Reverse lights housed in boot trim rail
Air conditioning	Combined heater and air conditioning unit mounted in fascia. Air blend system delivers air automatically to the required temperature to windscreen fascia and footwells
	Multi-directional outlets in outer ends of the fascia and central vent
	Automatic and manual override fan control with four speeds. Through-flow ventilation via rear quarter panel outlets

External dimensions

Overall length	4.87m (191.72in)
Overall height	1.26m (49.65in)
Overall width	1.79m (70.60in)
Wheelbase	2.59m (102.00in)
Front track	1.47m (58.00in)
Rear track	1.49m (58.60in)
Ground clearance	140mm (5.50in)

Interior dimensions

Front headroom	914mm (36.0in)
Rear headroom	826mm (32.5in)
Maximum width - front	1422mm (56.0in)
Maximum width - rear	1346mm (53.0in)
Seat squab to brake pedal	521-362mm (20.5-14.25in)

Luggage compartment

Maximum height	565mm (22.25in)
Maximum depth	572mm (22.5in)
Maximum width	991mm (39.0in)
Capacity	0.43m^3 (15 cu ft)

Weight	1687kg (3710lb)
	Weight includes automatic transmission, automatic air conditioning, energy absorbing bumpers, door side intrusion members, electric windows, central door locking, radio with electric aerial but less fuel

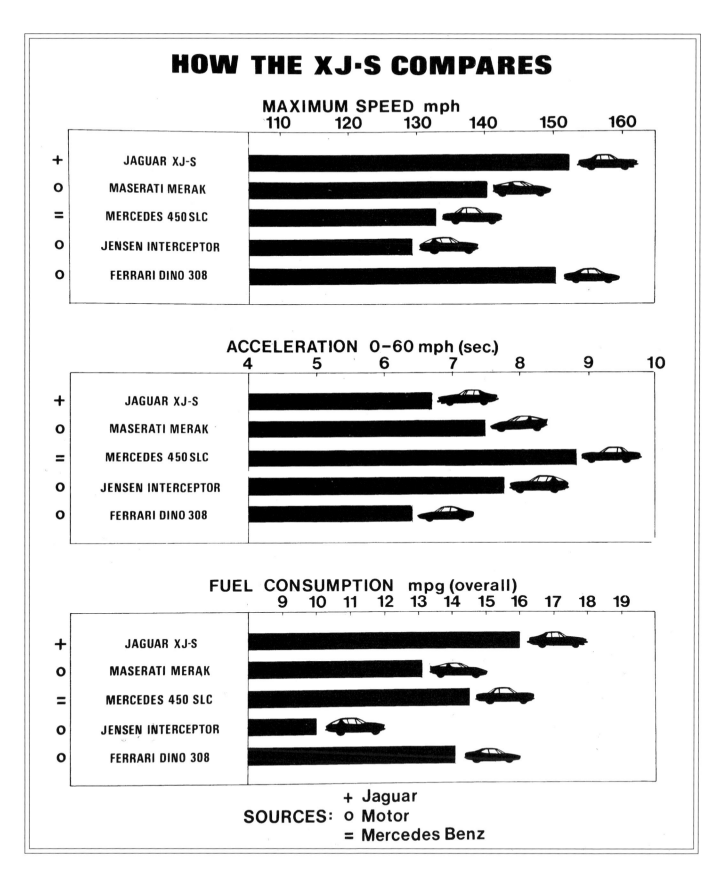

HOW THE XJ·S COMPARES

MAXIMUM SPEED mph

110 120 130 140 150 160

+ JAGUAR XJ-S
o MASERATI MERAK
= MERCEDES 450 SLC
o JENSEN INTERCEPTOR
o FERRARI DINO 308

ACCELERATION 0–60 mph (sec.)

4 5 6 7 8 9 10

+ JAGUAR XJ-S
o MASERATI MERAK
= MERCEDES 450 SLC
o JENSEN INTERCEPTOR
o FERRARI DINO 308

FUEL CONSUMPTION mpg (overall)

9 10 11 12 13 14 15 16 17 18 19

+ JAGUAR XJ-S
o MASERATI MERAK
= MERCEDES 450 SLC
o JENSEN INTERCEPTOR
o FERRARI DINO 308

SOURCES: + Jaguar
o Motor
= Mercedes Benz

Design highlights

Having described the leading features of the XJ-S in great detail, a few more sheets in the press pack described some of the car's background from a design and production angle, and also the way in which the new Jaguar was going to be sold. They read as follows:

A powerful answer to an instrumental problem

The problem facing Jaguar engineers was to match the XJ-S's performance with instruments that gave the driver instant information and prompted the fastest reactions. They wanted to eliminate the thought process that takes place between the driver registering that, for instance, the oil pressure was low and deciding what to do about it.

The answer was found in a power station. A Jaguar engineer noticed that important instruments were calibrated vertically with the mid-point as the normal operating condition. Needles that rose or dropped below the centre line were immediately noticeable.

This concept has been applied to the XJ-S fascia where four instruments - water, oil, fuel and volt indicators - are grouped together.

High performance from the XJ-S instruments

The instrumentation of the XJ-S has been developed to match the car's high performance with a comprehensive warning system that gives instant information and positively attracts the driver's attention in the event of a failure.

The mechanical and safety functions are monitored through a bank of lights which are colour-keyed red or amber. Red lights indicate major faults such as brake failure, or loss of oil pressure. Amber lights cover secondary faults like a side or brake bulb failure, or unfastened seat belts.

If there is a failure, a large red or amber light appears - which is known as an 'attention-getter' because the driver cannot fail to see it. If the red light comes on the driver should stop immediately and investigate the cause, while an amber light indicates that he should pull in as soon as possible. A second red or amber light pin-points the cause of the trouble.

The instruments indicating fuel level, water temperature, oil pressure and battery condition are of a type new to the British motor industry. Known as vertical 'air-cored' units, they work by a variation in a magnetic field between three opposing coils. This type of instrument gives a high degree of accuracy and reliability.

New safety bumpers - first in Europe

A bumper design new to Europe has been adopted by Jaguar to protect the XJ-S and its occupants in the event of low speed collision. The bumpers - both front and rear - are mounted on 'Menasco' struts which work on the same principle as telescopic shock absorbers. In collisions of up to 5mph, the struts absorb the impact effortlessly.

The secret of the Menasco strut is the use of silicone wax as the hydraulic medium. It has properties which en-

able it to both soak up the energy created by the collision and then to force the strut back to its original position. On impact the wax is forced by the piston through small ports in the chamber. After impact the wax returns to the chamber, forcing the piston and bumper back into position.

Although developed originally to meet American safety regulations, the bumpers will be fitted to all XJ-S models, wherever they are sold.

A new track for the XJ-S

The XJ-S has an assembly line all to itself at the Jaguar factory in Coventry. The new 2000ft track, built alongside the two lines for the XJ saloons and coupés, is part of a £6.5 million investment for developing and producing the car.

New tooling and press shop equipment has been installed at British Leyland's Castle Bromwich plant, where the XJ-S bodies are made and delivered to Jaguar already primed.

The new track has been designed to give maximum working space around the cars, and underneath them, with an elevated area at the beginning, dropping to ground level for fitting trim and a sunken section near the end for final underfloor work.

The XJ-S is also under close quality control scrutiny. With the new track a new method of zone inspection has been introduced. The track is divided up into 12 zones, and after each zone there is a quality control check. Any work that has to be done is carried out on the spot and not at the end of the assembly process.

The XJ-S at the time of launch. It was nearly christened the XK-F (after all, the Americans referred to the E-type as the XK-E), but rightfully, as the project developed, a name was chosen to distance the new GT from the legendary sports car. The Jaguar Le Mans was eventually rejected in favour of the XJ-S appellation, mainly because Pontiac owned the rights to the Le Mans title.

In this way, quality is continually monitored - and following Jaguar tradition, the XJ-S is subject to a rigid standing order at the factory: the sixth and final coat of paint is not applied until each car has been thoroughly road-tested and declared mechanically correct.

Selling the XJ-S worldwide

Sales of the XJ-S in its major market - North America - could total more than £26 million in the first year, with another £4 million from other export markets, says Keith Hopkins, Sales and Marketing Director of British Leyland Cars.

The United States and Canada have always been Jaguar's most successful markets. About 75% of all XJ-S cars built will be exported across the Atlantic, where the model's style and very competitive price will make it a big dollar earner for Britain. With produc-

tion set initially at about 60 cars a week - around 3000 a year - the balance will be allocated between the home market and the rest of Europe, with some cars for Australia. [As a matter of interest, XJ saloon production stood at about 750 cars per week at this time, with around 80% of these being six-cylinder models.]

The XJ-S is being announced in North America, Britain and European countries where the Jaguar XJ saloons are strong sellers - particularly France, Belgium, Germany, Switzerland, Italy and Holland - on the same day.

By sharing many components with the saloon range, the XJ-S will offer exclusiveness of style and low production volume, backed by a worldwide service and parts network.

"It is expensive compared with previous Jaguars, but is extremely good value compared with its exotic com-

petitors. Like all Jaguars ever built it offers much more at a lower price," says Keith Hopkins.

"Jaguar has never followed fashion. Fashions change but flair is timeless. This is a car that is equally at home on Sunset Boulevard, the Champs-Elysees and the Via Veneto. The styling is typical of Jaguar - modern yet restrained. It has an aesthetically pleasing individuality with a wide customer appeal. This has always been the basis of Jaguar styling and the XJ-S takes this into new areas.

"We are announcing the XJ-S on the eve of the Frankfurt Motor Show with the deliberate intention of letting the opposition see that we are really out to capture a large slice of the market. It will be the spearhead and the figurehead for the new British Leyland cars."

JAGUAR

3

THE FIRST GENERATION

"Jaguar's new car has been rumored for at least a decade, and there was almost a different rumor for every one of those years. Mid-engine roadburners to blow the doors off the Italian rockets and make the same mark left by the first XK120s and E-types. Super sedans, able to humiliate Mercedes-Benz. Refined touring cars, built of existing components. Racers, designed for the Mulsanne and adapted to satisfy an insatiably hungry market. The XJ-S emerges finally as the car with a little bit of all those." - Road Test, February 1976.

The Frankfurt Show is traditionally held in September. The theme of the 1973 show was "Into the future with the automobile," but the events that followed the Yom Kippur war left a big question mark over the future of the car *per se*. As can be seen from Chapter One, Jaguar was already faced with a lot of problems - some political, some due to in-fighting caused by the British Leyland regime, and some totally out of their control, such as the oil crisis sparked off by the Arab-Israeli conflict.

Anyway, the theme at the 1975 Frankfurt Show (which ran from 11 to 21 September) was "Live better with the automobile." The new Jaguar coupé could easily overcome its controversial styling, but it was pretty hard to justify a car the size of the XJ-S with 2+2 seating and a massive thirst for fuel in a time of cutbacks. What's more, at £8900 (well over twice what the 2+2 E-type had cost in its final year), pricing was also grandiose by Browns Lane standards.

There was very little in the way of exotic machinery at the German event. The Porsche 911 Turbo made its first appearance at Frankfurt, and Mercedes displayed its 450SEL 6.9, so it wasn't all economy cars or warmed-up versions of existing models, though it would be fair to say that there wasn't too much to get excited about.

On the British Leyland stand, the revolving red XJ-S was surrounded by examples of the latest Triumph TR - the wedge-shaped TR7, announced in May 1975 as "the shape of things to come" - and the Austin Princess. There was also a cutaway V12 engine to

An early car at speed on a British motorway. Unfortunately, speed limits imposed by the Labour Government in 1965 restricted the XJ-S driver to using less than half the car's power.

53

JAGUAR XJS

Various views from the first XJ-S catalogue - pure 1970s. Both British and French registered cars were used for the photography.

boost Jaguar content, but the Leyland hierarchy was certainly carrying out its plan to bring all BL products under the same umbrella.

It was sad to see the British launch at Earls Court in October, the Jaguar GT sharing a stand with the newly-announced Mk II Morris Marina and the ghastly Allegro Estate. As *Classic Cars* said: "Big L is making a great effort to appear dynamic and go-ahead in this Diamond Jubilee show, and is lumping all its offerings together under one 'Leyland Cars' banner instead of under individual marque names - an honest move reflecting internal policy, but one which will no doubt annoy (or perhaps sadden) Jaguar enthusiasts, to name but a single group of die-hards."

It is interesting to note, however, that even during the darkest days of British Leyland, Jaguar's management would not allow the BL badge to be fitted to the outside of any of the cars that left Browns Lane, despite the fact that every other marque in the BL empire was forced to display the company's small square badge.

BL Director, Keith Hopkins, said at Earls Court: "The 12-cylinder E-type roadster marks the last of the line of that type of car which started with the XK in 1948 and continued with the E-type in 1961. These were trendsetters in their way and we hope that the XJ-S will also be a trendsetter."

All Jaguar could do now was wait and see ...

Initial reactions in Europe

Expectations were naturally high, but, thankfully, at least overall, few were disappointed. The *Autocar* said: "An XJ12 driver put into the driving seat of the XJ-S would certainly notice more performance, a slight and valued improvement in the saloon's already superb handling, and different visibility - the XJ-S tends towards American and Italian fashion in super-fast cars where the faster the car, the less rearward view is deemed necessary - but he would find it more refined if anything, and as gentle to drive.

"Whatever one thinks of the appearance of the XJ-S - for what personal prejudices are worth, not all of us find this Jaguar as immediately beautiful as several of its predecessors - to drive the XJ-S, even for an after-noon, is to admire it very much."

Paul Skilleter, the respected Jaguar historian, once wrote: "The XJ-S was usefully quicker and tauter-handling than the 12-cylinder Jaguar saloons, yet retained their, by now legendary, disdain for poor road surfaces and almost complete imperviousness to mechanical, road and (to a slightly lesser extent) wind noise."

"This is a stupendously good motor car, a credit to British Leyland, Jaguar, and a much-needed fillip for the British motor industry as a whole ... In its diverse performance attributes the XJ-S is unequalled. With more attention to detail finish and appointments and some styling revisions it could so easily be the best car in the world," said Clive Richardson of *MotorSport*.

Overall, *Motor* was also complimentary: "Tank or Supercar? A bit of both. It's large, heavy, thirsty and cramped in the back. It's also superbly engineered, sensationally quick, very refined and magnificent to drive - a combination of qualities no other car we've driven can match at the price."

John Bolster wrote in *Autosport*, "A car of this price and performance is scarcely a logical buy, but as long as there is a demand for such exotica, Britain should endeavour to supply it. I've driven the German and Italian equivalents, and very nice they are too, but nobody who has tried them all could be in any possible doubt. The V12 from Coventry is the one, and if they can make them all as well as they built the car they lent me, Britain has a world-beater."

The Lord Mayor of Birmingham, England, trying the new car for size at the Birmingham Motoring Festival in September 1975. The vehicle was supplied by P.J. Evans, a local BL dealership.

Even *Classic Cars* noted: "It is almost impossible to fault Jaguar's XJ-S, and to drive one on a long journey is still a memorable experience in one's motoring life, no matter what one is used to driving. In the XJ-S we have a vehicle which will outrun virtually any sports car, and yet match the world's greatest luxury saloons for comfort, ride quality and silence. A quite amazing blend of contradictions and a unique technological achievement by Jaguar's small group of engineers."

Autocar was extremely impressed by the refinement: "It is sensationally quiet - the engine was virtually inaudible even at speeds of around 60mph, with a subdued hum when you used acceleration to the full, very low bump-thump, hardly any rolling noise except when badly provoked by harsh concrete, and very little wind noise except at above an indicated 130mph ..." This enthusiasm was later echoed in *Motor*: "Jaguar are masters at suppressing noise, vibration and harshness and the XJ-S is a brilliant example of their technological achievement."

Perhaps the most commonly praised aspect of the XJ-S was its V12

power-unit. The smooth and willing flow of power was totally captivating, and, unlike so many high-performance engines, it was available throughout the rev range. "There is no temperament, no fuss or bother, no abrupt change, no 'coming on the cam,' in fact, none of the characteristics which are often the hallmarks of cars with equal or better performance but with more highly tuned and stressed engines. The engine is so flexible that it is possible to pull away cleanly and vigorously from a walking pace in top," said *Motor*. Indeed, one magazine even managed to carry out a test that took the car from standstill to over 140mph in top gear without touching the clutch.

MotorSport took a car to Munich and recounted the following: "Letting the XJ-S have its head from rest takes it to 50mph in first gear, 84mph in second and 116 in third, with 60mph coming up in 6.7 seconds and 100 in 16.7 seconds. The ratios are lower than need be for the 294lbft of torque; second will cope with most standing start conditions - well, so too will top. At Britain's 70mph legal maximum, our passage was silent save for some

roar from the 205/70 VR15 Dunlop SP Super tyres. Some wind hiss emanated from a faulty nearside door seal, changing to a sudden roar at exactly 130mph. But engine noise was little more than a whirr, even up to the 6500rpm red line. 100mph is roughly 4000rpm in top, nice and lazy, but even on this gearing the engine felt to be so much on top of its work that the overdrive, which the handbook confirms will become optional in production, might have been a good economic asset. Nevertheless, our roughly 100mph average from Calais to Cologne produced 13.57mpg."

In ideal conditions, the manual XJ-S covered 0-60 in 6.7 seconds, going on to achieve 100mph in just 16.2. Mid-range performance was even more spectacular, with 50 to 70mph taking 6.9 seconds, and 70 to 90 only 7.2; a top speed of 154mph was recorded. This compared well with the Ferrari 365GT4 2+2 and Mercedes 450SL, and was only a fraction off the blistering pace of the Aston Martin V8 and Maserati Bora.

However, even with supercar performance on tap, given the economic climate of the time, fuel consumption also had to be considered. As noted above, *MotorSport* got 13.6mpg, while *Motor* returned 12.8mpg after some spirited driving with the manual car, although an average of 15 to 16mpg was perhaps more in keeping with everyday use - not bad for a 5.3 litre car at the end of the day.

As for the transmission, most of the early tests refer only to the manual gearbox. *Autocar* noted in September 1975: "Something XJ5.3 owners are

XKSS XK120 XK140 XK150
C TYPE D TYPE
LE MANS 1951, 1953, 1955, 1956, 1957.
E TYPE XJ6 XJ12

Some claim heritage. Others have truly earned it.

Almost every new Jaguar has been a breakthrough in design or performance. Or both.

The XJS is no exception. It has been designed to be the definitive Jaguar.

Which makes it, in almost every way, the definitive high-performance luxury car.

More than five years and £5 million went into its design and production.

With the result that no other car currently made offers a comparable combination of safety, luxury and performance. At anywhere near the price. (£8,900. 19p. inc. VAT, Car Tax and front seat belts, but not delivery charges or number plates.)

The list of technical innovations in the XJS is impressive. As is the list of luxury and safety items which are fitted as standard equipment.

And it goes without saying that the performance is startling.

Zero to sixty takes under seven seconds.

The top speed, where permissible, is in the region of 150 mph.

Yet this level of performance is achieved in levels of silence and safety that will astonish and delight you. As will the mpg figures.

Admittedly, because of the export demands for the XJS, you may have to wait a while before joining the elite few who own one.

But the wait will be well worthwhile. So let patience bring its reward. The Jaguar XJS.

The car everyone dreams of. But very, very few can ever own.

🐆 Jaguar

From Leyland Cars. With Supercover.

58

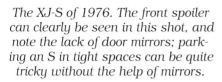

Left: Following the famous 'Black Day for Modena' advert, Jaguar issued this piece of advertising to declare the car's impeccable bloodline.

XJ-S production. Chassis numbers, incidentally, started at 2W-1001 for rhd cars and 2W-50001 for lhd models.

The XJ-S of 1976. The front spoiler can clearly be seen in this shot, and note the lack of door mirrors; parking an S in tight spaces can be quite tricky without the help of mirrors.

not able to appreciate to the full, because that model is only obtainable with automatic transmission, is the truly extraordinary flexibility of the V12; formerly only owners of manual box E-types could realize the engine's full range. Happily, XJ-S buyers will be allowed the pleasure - and better efficiency - of a normal gearbox, Jaguar's four-speed, with the revised ratios introduced when the XJ3.4 was announced - first gear being set lower, at 3.238:1 instead of 2.933. A 3.07:1 final drive is used with a Salisbury Powr-Lok limited-slip differential, as on the latest XJ5.3 giving overall gearing of 24.7mph per 1000rpm [the same as that for an automatic XJ-S, of course].

A typically English setting for a typically British Grand Tourer. (Courtesy Paul Skilleter)

At the engine's peak power speed, this will correspond to 135mph, so that if Jaguar's claims of a maximum speed 'of 150mph plus' are justified - and driving impressions suggest they will be - then the car is, at the moment, undergeared."

There is always a contradiction in views regarding the S and, sure enough, *Motor* thought the manual gearbox wasn't such a pleasure: "One of the less endearing aspects of the XJ-S is the transmission. The clutch is a little heavy and unprogressive, and initially there was some stickiness, though this later seemed to clear up. The change is stiff and notchy if you rush it, so it pays to ease the lever around rather than snatch at it. A distinctly audible whine in first and second gear detracts from the overall refinement of the car."

Naturally, compared to the manual transmission, the automatic was less spritely off the mark (0-60 came up in 7.5 seconds), and generally less sporting in its nature. For this reason, most of the press cars in the fleet were equipped with the four-speed manual box, but the automatic gearbox was perfectly acceptable, and sales figures later proved this. One of the few things mentioned about the automatic gearbox was the lack of a detent between neutral and drive, which has caught out many drivers over the years, including the author, despite being a Jaguar owner for well over a decade.

The steering, although still a little on the light side, was almost universally praised, but the braking received some mixed reactions. While *Motor* gave the brakes top marks in two separate tests, *MotorSport* had unexpected problems, stating that: "all was not well with the XJ-S, which pulled halfway across the road under braking. A return to Jaguar gave new pads and re-routing of the radiator overflow pipe (already done on later production cars) to prevent it pumping anti-freeze mixture over the nearside caliper."

Apart from the occasional tendency to "float" slightly over undulations (most testers agreed it was a small price to pay in a car that was otherwise so composed), the ride and handling were pretty much exemplary. "Lifting off or braking in a corner simply tightens the line without any drama," said *Motor* during its test of February 1976.

Autosport was equally taken by the car's manners: "Near-neutral handling and exceptional traction are its outstanding characteristics. It is ultimately possible to make the rear end

60

The XJ-S with the mid-engined XJ13 racer, featured here in an H.R. Owen publicity shot. As well as dealing in a number of the world's most prestigious marques, H.R. Owen still hold a Jaguar franchise.

break away on a test circuit, but only by driving much harder than would be in any way reasonable on the road. The behaviour continues to be safe and predictable on wet surfaces and the steering is very much better than the rather over-assisted set-up on other Jaguars, though there is still not much feel of the road."

Inside: "the shapely seats are comfortable and provide the right sort of support in the right places," said *Motor*, but the small rear seats were criticised. Autosport noted: "The XJ-S is a wide car, giving superb comfort for two people. The rear seats are too cramped for adults on long journeys, but they are ample for short trips and, of course, perfect for children. The driving position is comfortable and the steering column length is adjustable, though a short driver may have difficulty fully depressing the clutch pedal. Although there are cars with heavier clutch operation, this one may become tiring at the traffic lights. I do feel that all powerful cars should have servo-assisted clutch pedals, as the Packard did before the war, if only to match the lightness of the brake application. "

The dashboard was "functional and luxurious" as one road tester put

it, although most pointed out that the silver-painted rims on the dials looked cheap. Heating and ventilation was classed as average (the standard air conditioning unit was the same Delanair system as found in the Series 2 saloons), although luggage space was a real bonus. The boot is exceptionally big for a GT car, and will hold a surprising amount - the author once managed to fit in a tumble drier, in its box, and still close the bootlid! There was also an excellent toolkit.

It should be noted, though, that for such an expensive car - the price was already up to £9527 within a few months of the launch, at a time when even the Daimler Double Six VDP was only £8045 - there were a number of omissions, considering the luxury GT specifications. There was no windscreen wiper delay facility, a petrol filler lock was classed as optional, and a headlamp wash/wipe system wasn't available at all. There were no kph markings on the speedometer, no lighting in the glovebox or under the bonnet, the standard aerial was manually-operated, the gaiter on the manual gearbox was plastic, and when outside mirrors were fitted, they couldn't be adjusted from the driver's seat. "Poor

detailing," noted *MotorSport,* in view of the price.

There was also an unexpected problem with the German market, with the authorities considering the "flying buttresses" dangerous, describing them as sharp edges. Jaguar wasn't the only manufacturer to fall foul of this ruling, though, as the Lancia Beta Montecarlo was also banned. Even the rear spoiler on the Porsche 911 Carrera was disliked; fortunately, common sense did prevail in the end.

However, on television, actor Ian Ogilvy could be seen on a weekly basis driving a white example in the smash series *Return of the Saint*. Jaguar officials had originally refused to lend an E-type to the producers of the first Roger Moore series (he ended up using a Volvo coupé instead), but now they were more than glad of the free publicity.

Another interesting aside was the Nepro 'XJ-S' quartz watch. Introduced during the early part of 1976, the XJ-S rear window and flying buttresses inspired Nepro, the Swiss watchmaker, to introduce the 'XJ-S' quartz LED timepiece. Using a Swiss-registered car in the advertising, it was quite a novel and elegant idea.

Despite the low-key launch, the

Ex-F1 World Champion, Phil Hill, was a strong supporter of Jaguar's new GT. This picture was supplied to the US press.

company furnished every dealer with a car and the initial rush to get the XJ-S was undoubtedly encouraging; after the announcement, it was reported that there was a nine month waiting list for the new model.

Another piece of encouragement came via *Motor* magazine, which tested the manual S against an automatic Mercedes 450SLC in May 1976. It concluded: "It's a dead heat on points. So, how do you choose between two such magnificent motor cars? For some, of course, the clinching factor will be price: the Jaguar is £2223 cheaper, after all, for others the decision may be more abitrary. If so, for us it would still be the Jaguar, that uncanny, unrivalled refinement tipping the balance in Coventry's direction, but only just."

The new car in America

Although the event was officially for the Triumph TR7, the XJ-S was briefly introduced to dealers in the US in January 1975 at a meeting held in picturesque Boca Raton, on Florida's east coast. Shortly afterwards, pictures appeared in a newspaper, and then *Road & Track* carried a two-page article with spy shots of the new Jaguar and drawings by P.J. Prescott. At this stage, although the press embargo had been broken, early signs were very good.

Then, in the second week of September, the official west coast launch took place at the Long Beach Hilton in California, with that for the east coast being held in the heart of the American motor industry, the Detroit Yacht Club, one day earlier. The team in the States (Jaguar Cars Inc. was established in

Pages from the first American catalogue. Note the additional side markers on the front wing and the small badge underneath the fuel lid warning owners to use unleaded fuel only.

The striking cover of the first American catalogue. In America, the XJ-S was often referred to as the S-type, a name which was even used in some of the early brochures and advertising published in the States.

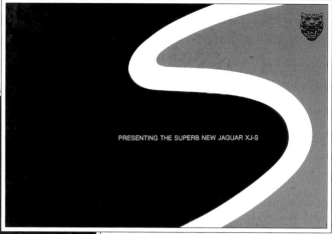

PRESENTING THE SUPERB NEW JAGUAR XJ-S

Through the leather-covered rim of the Jaguar XJ-S steering wheel, which is adjustable for distance, the warning lights along the top of the gauge nacelle are in easy view. Two bright lights, red or amber depending upon the degree of urgency, reinforce the messages that are explained in detail by the smaller lights.

The many comfort provisions of the S-type are monitored from the central console. Among them are the controls for the AM/FM stereo radio and automatic transmission, both of which are standard equipment. The clock is centered between press-button switches for lights and the electrically-heated rear window. Flanking a central outlet are two of the four adjustable vents that admit face-level ventilating air.

Seeking a new and effective way to convey vital engine information to the driver, Jaguar engineers fitted the S-type with electro magnetic gauges. All is well under the hood if a glance shows that the needles of these vertical-reading gauges are aligned at mid-dial level.

All the functions of the standard-equipment integrated heating and air-conditioning system are selected by two dial controls. One chooses the mode of operation and the other sets the interior temperature, which may be maintained by the automatic air-mixing system. Face-level air is kept cooler than that to the footwells to help the occupants stay fresh and alert.

1954, although it was known as Jaguar Cars North American Corp. in the early years), would have an important role to play in deciding the fate of the XJ-S.

It was, therefore, surprising that only two cars were available in the States at the time: a red one for Detroit and a silver one for the west coast launch. Phil Hill, the former-F1 World Champion, had given the S his endorsement and was present at Long Beach (as well as the XJ-S event, it was the Long Beach Grand Prix weekend). Jonathon Heynes, the son of Jaguar Director, Bill Heynes, was present to ensure the vehicle ran well, and afterwards Hill actually bought the first silver press car.

Ultimately, however, as in Britain, reactions regarding the $19,000 car were mixed, to say the least. Mike Knepper, a part-time racing driver who wrote for *Road & Track* at the time, questioned the very validity of such a vehicle: "With the XJ-S, Jaguar has ventured into unfamiliar territory and will be courting an unfamiliar segment of the market. Although Jag marketing types talk of their new creation going head-to-head with the Mercedes 450SLC with a 'style and flair the 450SLC doesn't have,' it will also attract the attention of Merak, Urraco and Dino buyers. Jaguar apparently doesn't intend for their new car to keep such heady company, but if you put an exotic V12 engine in an exotic 2+2 GT body, what can you expect? I think

American advertising from the early part of 1977. To comply with Federal lighting regulations, cars destined for the North American market had four separate GEC tungsten headlights instead of the Cibie halogen units found on European models. Although a similar set-up was appraised for Europe, the idea was rejected by Jaguar management.

the styling - which is basically as originally conceived in 1970 - is, well, not good."

On the other hand, in the January 1976 issue, *Car & Driver* carried the summary: "The XJ-S is a dark and mysterious product of England's tortured auto industry, fantastically over-qualified for today's driving conditions - a rich man's car for an opulent age that might have been but now clearly never will be. That is reason enough to want one."

A couple of years later, *Motor Trend* came out with a statement that was almost exactly the opposite to that made by Knepper: "In spite of its flaws, the Jaguar XJ-S remains one of the finest mass-produced cars in the world. It has personality, presence and a trait that is becoming rare in automobiles; character."

As the British *Autosport* magazine pointed out, an "important feature of the engine is its clean exhaust, the vertical valves and Heron-type combustion chambers having better anti-pollution characteristics than the classical inclined valves of earlier Jaguar models. For a car aimed at the export market, this is vital."

However, Federal specification cars had an 8.0:1 compression ratio and an EGR system, thus reducing power to 250bhp; with California's additional anti-smog equipment, this further reduced maximum power output to just 244bhp at 5250rpm (peak torque was quoted as being 269lbft at 4500rpm). Another change for the American market was the gearing. Although the gearbox ratios were the

there's no question the front-engine XJ-S will be able to more than hold its own in performance with those three mid-engine exotic cars. Its refinement really can't be matched by the Dino, Merak or Urraco and it will cost less than any of that trio. On paper, it's a strong threat. Maybe Jaguar has a double-threat in the XJ-S. But maybe not. The 450SLC and the aforementioned Italian trio represent two ends of the supra-expensive 'sporty' car

market: refinement and manners vs sensuality and performance. The Jaguar is a strange combination of both which just may not appeal to either segment of the market."

He summed up his article with the words: "I like the XJ-S, but my enthusiasm doesn't go much beyond that. In their attempt to create an ultra-refined, luxurious GT machine, Jaguar engineers have engineered out any personality the car might have had. And

same, the final drive was changed to 3.31:1. The performance was still more than adequate - with an automatic transmission, 0-60 was clocked at a shade over 7.5 seconds, whilst top speed was in the region of 140mph.

Refinement was again a key point. *Car & Driver* said: "A car that is silent and surefooted at 140 has to be even better at 55. The engine is quiet - uncannily so. You hear other sounds instead; the expansion of Freon in the air conditioner, a whine from the transmission, even the speedometer cable. It seems uncarlike. Exotic. But when you toe into the accelerator, the engine takes over with authority. Air moans as it enters the twin cleaner boxes. The car deliberately gathers itself up before it moves forward. It's very quick - 90.3mph in the quarter-mile - but it denies you the thrill of speed. It never seems to go fast, even when the speedometer needle is pointing out three-digit numbers. It's too composed to seem fast.

"The suspension is amazing. It is taut but not hard, and the geometry is accurate. The car seems immune to the laws of physics. It makes you think you could never lose it - that it would compensate automatically for your blunders."

Even the racing driver, Bob Tullius, who has campaigned the V12 coupé, was impressed with the road car's composure. He said: "Even though it's a street car, this car does what it's supposed to do. It handles much like the race car if you can interpolate the difference between the need for the race car to be stiff and the street car to

View from the pack.

The Jaguar S-Type.
Perhaps the best-handling four-passenger car in all the world.

Being in the vanguard is always an honor. Leading it is an uncommon distinction. Jaguar has earned this distinction in forty years of building superb high-performance motorcars. And, while Jaguars have always been known for their extraordinary handling ability, the S-type stands out dramatically, even among Jaguars.

Car and Driver says: "The suspension is amazing. It is taut but not hard, and the geometry is accurate. The car seems immune to the laws of physics."

The superb handling of the S-type is matched by its depth of quiet luxury. *Road & Track* says:

"The emphasis is on refinement, complete silence, luxury, comfort and general opulence, and it will run the pants off a 450 SLC."

The S-type is powered by the smooth, silent but amazingly powerful, electronically fuel-injected V-12.

The S-type is, indeed, the ultimate Jaguar—the best-handling Jaguar ever built. For the name of the Jaguar dealer nearest you, call these toll-free numbers: (800) 447-4700, or in Illinois, (800) 322-4400. British Leyland Motors Inc., Leonia, New Jersey 07605.

be soft. The only real difference between it and the race car is the throttle application. With the street car you can stand on the throttle and it won't spin the wheels and send you roaring into the woods somewhere. When the car goes into a corner it tends to understeer just a bit until you get the throttle on and then it gets up on its haunches and does a little broad slide or the tail hangs out a little. It's exactly

like the race car only softer and gentler" - a testament not only to the talent in Jaguar's engineering department, but also to Sayer's skill in mastering aerodynamic form.

Equally, *Road Test* was left in awe of the Coventry machine: "The eerie silence, the total dignity of the car at speeds well over twice the national limit, could get the careless driver a lot of long conversations with the law. Not

Further proof of British Leyland's restricted advertising budget; Japan's brochure (for the entire XJ series) also had photography from the first American catalogue. This lack of investment in a market hungry for supercars, allied to a bad reliability record, severely limited Japanese sales.

that you necessarily try to ignore speed limits. It's just that the XJ-S is, to use an incredibly worn cliche, as smooth and effortless at 90 and more as most other cars try to be at 55, and so by accident rather than intent you find yourself continually above legal speeds.

"We know, in our hearts, that this car is a quality item, probably the best car Jaguar has built. But our only experience with it in this country has revealed shortcomings. Certainly, the individual owner starting with a new car and giving it proper maintenance should find it a vast improvement over previous Jaguars."

Road & Track tried the same car. "To sum up the XJ-S, one can say that it is a superb automobile in the exotic, luxury class and, with the exception of the automatic transmission, does exactly what it is designed to do. However, it has two drawbacks in performance and ease of purchase for those of us accustomed to earlier Jaguars. The first is that it does not replace the E-type and the second is that, unlike previous Jaguars, whether or not you want to own one is not entirely up to

you, it's entirely up to your bank."

Sadly, for most, especially in view of the price, the interior was a real disappointment. *Car & Driver* pointed out: "Those who are attracted to the XJ sedan for its opulent interior will be somewhat disappointed by the XJ-S. For one thing, there is no wood. Really, this shouldn't be a surprise. Sporting Jaguars do not have wood dashboards. Only the sedans. But it is so well done in the sedan, such a perfect complement to the Connolly hides on the seats, that we've come to think of it as a trademark of Jaguar cars and are therefore disappointed to find all-black, vinyl-covered padding on the panel of the XJ-S. The vacuum-formed plastic instrument cluster, a piece that is pure Pinto in design and execution, only heightens the loss. The seat facings do remain leather (the sides are vinyl), however, and a high quality carpet is used on the floor and in the trunk."

There was also the question of reliability - trouble with the early fuel-injection and electrical systems, in particular, and suspect build quality. Paintwork was often poor, and it was

later admitted that the second silver press car was repainted in the States to avoid embarrassment. Even dealers were losing faith in the Big Cats. Add in the fuel crisis that gripped America at the time, and it was obvious that the team in New Jersey had an uphill battle on its hands.

Only 287 XJ-Ss were sold in the States in 1975, rising to a 1970s peak of 1365 the following year, by which time the basic price had risen to $20,250. For the first two years, standard colours included Silver Grey, Old English White, Carriage Brown, Regency Red, British Racing Green, Squadron Blue, Dark Blue, Greensand, Fern Grey, Yellow Gold and Signal Red.

A new gearbox

From April 1977, the GM400 automatic gearbox replaced the aging Borg-Warner unit. At the same time, a number of detail changes brought the car up to the standard expected for the price, which was now in excess of £13,000. Door mirrors could now be adjusted from inside the car.

Advertising from Coombs of Guildford in the UK. John Coombs had done so much to enhance the image of the Mk II through his racing exploits, and the number plate featured on this XJ-S has been passed down from one of the most famous of all Coombs' competition cars.

At the end of May 1977, *Autocar* noted: "The XJ-S was originally offered with the Borg-Warner Model 12 automatic as used in all XJ4.2 and 5.3 models for some time. As we now know, it was Jaguar's intention to replace the Model 12 with the General Motors THM400, which Jaguar considered a superior transmission in many ways. For the first year or so of production, therefore, tests of automatic XJ-Ss were conspicuous by their absence.

"There are no changes to the XJ-S other than those physically demanded by the automatic transmission. The final drive ratio, and hence the overall gearing in top, remains the same at 3.07:1; so direct performance comparisons are valid."

Autocar found the new transmission (with ratios of 2.485, 1.485 and 1.00:1) gave "rapid yet smooth changes," but performance was definitely blunted when compared to the manual car. The best that could be mustered from the test car was 143mph, while the 0-60 and 0-100 stages were covered in 7.5 and 18.4 seconds respectively. Fuel consumption was an average of 14.0mpg.

These figures compared favourably with those of the £17,960 Maserati Khamsin, which was similar in most respects but had a much lower top speed, and those of the slower but slightly more expensive Mercedes-Benz 450SLC. Only the Aston Martin V8 could provide better performance, but it cost £3400 more - at that time, not far off the price of a 1.6 litre Alfa Romeo Alfetta GT.

But figures alone do not tell the whole story. As *Autocar* noted: "The use of tyres of 'only' 205-section is a well-judged compromise, which takes into account the need for reassuring stability (especially over roads of varying camber) as well as the excellent steering feel and reasonable turning circle. As things are, the XJ-S is very high in the league table of handling and roadholding, though by no means at the very top. What is more important, except to the sporting purist, is that the handling is sufficiently good-natured that even a moderately skilled driver can use most of it. Certainly one would hardly credit that nearly 55% of the [1754 kg] kerb weight rests on the front wheels, for any understeer is hardly noticed."

It is interesting to note that the brakes again came under scrutiny: "The brakes themselves are extremely good in most respects, though with two minor reservations. One concerns their lightness. Our other doubt about the brakes came when achieving the best possible stop. It became clear that the back wheels tended to lock slightly prematurely, leading to some mild control problems in a panic stop. We also noted a degree of wheel judder when stopping very quickly."

However, overall, *Autocar* was very pleased with the car, summing up the article with the words: "There are few enough cars at any price which offer as much as the XJ-S. If one is pedantic enough to insist on 12-cylinders and automatic transmission, there are only two, one of which costs half as much again, and the other, almost twice as much. In other ways, the Jaguar finds more logical competition in cars like the Mercedes 450SLC, which is priced very close to it and offers more room though less sheer performance. Per-

The Jaguar S-type is strong. It is quick and agile in its response to any challenge on the road. That strength and agility comes from a unique source: the famous Jaguar electronically fuel-injected V-12 engine. An engine that is only 5.3 liters in displacement, yet develops an astonishing 244.4 horsepower at 5250 RPM. Jaguar engineers call it virtually indestructible.

To further confirm that strength, an S-type took five Category I victories and the Driver's Championship in 1977, its first Trans Am season.

To match its uncommon power with its handling, the S-type is fitted out with independent suspension all around, very precise power-assisted rack and pinion steering, four-wheel power disc brakes, and steel belted radial tires. In fact, the XJ-S may well be the best-handling four-passenger car in the world.

The sleekness of the S-type is not merely cosmetic. It was achieved through exhaustive wind tunnel tests. Its purpose: to give the XJ-S even greater stability at speed.

And the silence of Jaguar's S-type is golden. It is the result of great care and craftsmanship: thick rugs on the floor, rich Connolly leather seats, thermostatically-controlled heat and air conditioning, AM/FM stereo radio and tape system and so many other thoughtful and luxurious touches that there are no factory options available whatsoever.

The XJ-S. Strong, sleek and silent. Here is a car of such uncommon capabilities and luxuries, that it may well redefine your expectations of what a grand touring car can deliver.

For the name of the Jaguar dealer closest to you, call these numbers toll-free: (800) 447-4700, or, in Illinois, (800) 322-4400.

BRITISH LEYLAND MOTORS INC., LEONIA, NEW JERSEY 07605.

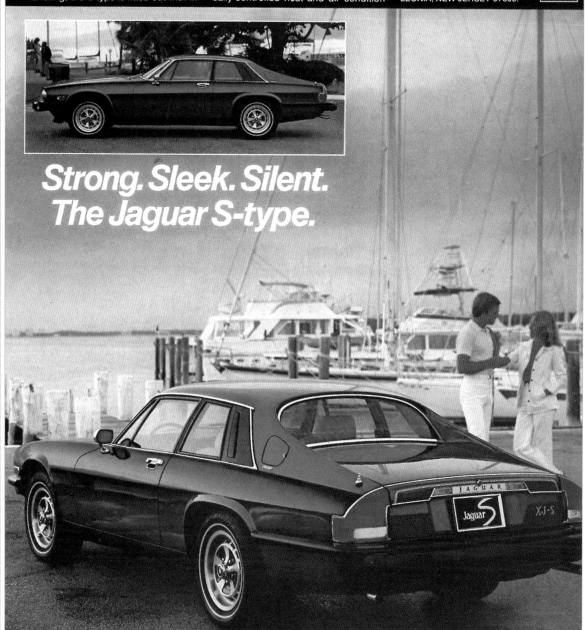

Strong. Sleek. Silent. The Jaguar S-type.

The 'Silver Jubilee XJ-S' at the 1977 New York Show. Graham Whitehead, then President of British Leyland Motors Inc., is on the left.

Unique interior of the 'Silver Jubilee XJ-S'.

haps the greatest asset of the XJ-S is that it can be whatever the owner wants it to be: one of the most capable and quick genuine GT cars, or a civilised, undemanding and incredibly refined carriage."

Other minor changes followed in autumn of that year, and included painting the lower panel on the bootlid in body colour instead of black, adding chrome finishers to the rear light infills, and giving the B-posts a black finish rather than chrome "to emphasize the horizontal lines of the car." Inside, the silver lines encircling the instruments, having been slated by the motoring press, were at last deleted.

Coachwork colours at this time included Old English White, Juniper Green, Carriage Brown, Moroccan Bronze, Regency Red, British Racing Green, Squadron Blue, Dark Blue, Yellow Gold and Signal Red.

At the 1977 New York Show, Jaguar displayed the 'Silver Jubilee XJ-S' - a special vehicle produced to celebrate HM Queen Elizabeth II's 25th year as Britain's monarch. Designed by Count Albrecht Goertz (who had penned the beautiful BMW 507 and was involved with manufacturers as diverse as Porsche and Datsun), it featured silver and gold coachwork with purple accents, and a special champagne gold and purple mock-suede interior. It was a good publicity stunt, but still, with the price now at almost $24,000, XJ-S sales were very slow in America, and only 869 cars found new owners in 1977.

XJ-S in the Antipodes

The XJ-S was due for release in Australia in the middle of 1977, but British Leyland had a pre-production car from the early part of 1976 in order to ensure the new model would comply with Australia's difficult Design Rules in time for the launch.

In the October 1977 issue, *Modern Motor* tested the automatic S, and was totally enthralled. In the magazine's five star summary, comfort, performance and quietness scored top marks, whilst all the other categories (handling, brakes and luggage capacity), were given four out of five stars. *Modern Motor* said: "Jaguar's decision to replace the Borg-Warner box with a General Motors unit seems to have been a wise one. The changes are smooth and precise, even when the big

For those few among us who are totally committed to excellence in all things.

British advertising from the autumn of 1978.

cat is being pushed hard, and while there is a manual override on the gearbox, using it is a rather pointless exercise because there seems to be limitless power for overtaking manouevres and superb brakes for slowing the beast down. The gate also proved to be far too stiff and vague for accurate manual changes.

"The seats themselves are really comfortable, both back and front, and have been carefully moulded to give

69

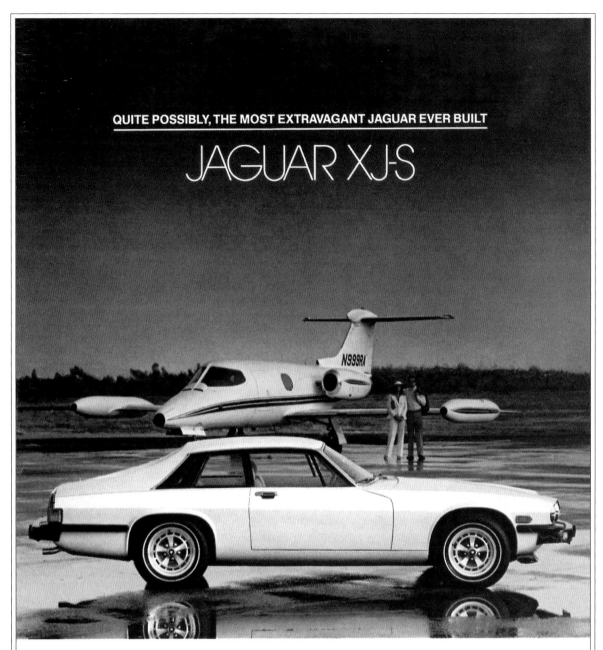

QUITE POSSIBLY, THE MOST EXTRAVAGANT JAGUAR EVER BUILT

JAGUAR XJ-S

Jaguar engineers and designers have always gone to great lengths in their pursuit of excellence. With the latest great Jaguar, the S-type, they've gone even further. It is extravagant in the degree to which it surpasses the limits of conventional luxury cars. For the XJ-S achieves, very gracefully, an almost magical blending of great luxury with out-of-the-ordinary handling and response. It may well be the best-handling four-passenger car in the world.

The S-type has fully-independent suspension, four-wheel power disc brakes, rack and pinion steering, and the velvet-smooth, immensely responsive electronically fuel-injected Jaguar V-12 engine. In its first Trans Am season, an S-type won five Category I victories.

Yet the S-type is surpassingly luxurious too: it is silent in motion and equipped with thoughtful accessories such as thermostatically-regulated heat and air conditioning, stereo AM/FM radio and tape system, automatic transmission and steel-belted radial tires. In fact, so completely equipped is the XJ-S that there are no factory options at all.

Perhaps *Road & Track* summed it up best when it said of the S-type: "The emphasis is on refinement, complete silence, luxury, comfort and general opulence, and it will run the pants off a 450SLC." Extravagant? Not for the Jaguar S-type. Drive it soon.

For the name of the Jaguar dealer nearest you, call these numbers toll-free: (800) 447-4700, or, in Illinois, (800) 322-4400.

BRITISH LEYLAND MOTORS INC., LEONIA, NEW JERSEY 07605.

70

good support over long distances. One great step forward in the area of improving driver comfort has been to put a leather glove on the steering wheel. But while the leather covering is an improvement, the wheel still needs to feel a bit beefier. The high beam/indicator/wiper/washer stalks on the steering column are easy enough to use but the operation of the windscreen wipers is totally out of keeping with the car. While they worked well, they were too small, and had a ridiculous 'park' phase, where they went into limbo after use before returning to a station that was far too high up the windscreen. On such a delightful and well engineered car as the XJ-S, this was a great disappointment, as we had come to expect greater things."

However, the lighting was highly praised, being "amongst the best we have come across, and sets standards that could well be adopted by other manufacturers."

In Australian trim, with a 9.0:1 compression ratio and 3.07:1 final drive, maximum power output was quoted at 212Kw at 5500rpm (283bhp), and 0-60mph was covered in 8.3 seconds; top speed was recorded as 149mph.

In summary, *Modern Motor* concluded that "the $32,750 Jaguar XJ-S is a truly magnificent piece of machinery. It is desirable in every way as a safe, high-powered touring car with superb handling and arguably the best interior comfort and sound levels of any GT on the market today."

A couple of years later, David Segal, reporting on the $38,290 automatic model for *Motor Manual*, said: "Although those who are familiar with Lamborghinis and Ferraris over similar distances tell me that the XJ-S isn't even in the same ball park, I've got to say that it provided me with the most comfortable and least tiring long distance jaunt I can remember."

Jaguar's saviour

Daimler buses were already being badged as Leylands, as were Guy trucks. Even worse, one could count on a yearly basis the famous names being lost to history, and everyone was wondering how long it would be before Jaguar went the same way. The only thing that seemed to save Jaguar from that awful fate was the Jaguar Engineering Department run by Bob Knight (who eventually became the Managing Director of the company), and a small band of enthusiasts led by people like Harold Adey.

At the end of 1977, Michael Edwardes was called in to oversee Leyland and, eventually, a number of departments that had moved to Cowley were brought back to Browns Lane. The fact that British Leyland had actually recognised the need for an MD of Jaguar Cars (the position went to Bob Knight in 1978) was indeed a good thing for the future of the company. Much more importantly, Jaguar was once again given almost total engineering freedom.

Described by *Autocar* as "a tough and intelligent people man," Edwardes observed that "great names like Rover, Austin, Morris, Jaguar and Land Rover, were being subordinated to a Leyland uniformity that was stifling enthusiasm and local pride." He proposed that marque names be revived and that the Leyland badge would only appear on commercial vehicles. Of the car producers, Austin-Morris would run as one concern, while Jaguar-Rover-Triumph (or JRT, the 'specialist' manufacturers) would run as another; Land Rover would become a separate entity altogether.

Death of the coupé

Having taken a year-and-a-half to enter the marketplace, it was quite a surprise to learn that the two-door versions of the XJ saloon were to be discontinued. In retrospect, although the cars were very attractive, relatively few were built - less than 10,500 between their announcement in September 1973 and the end of production in 1978. During those years, it was available in four forms - the XJ6C, the XJ12C and the two Daimler equivalents. With sales of less than 4500 cars a year, and the problems the coupé would have had complying with Federal side impact tests, it was fairly inevitable that it had to go - Jaguar was back to a two product line.

XJ-S Specials

The Ascot concept vehicle, designed and built by Bertone, made its debut at the 1977 Geneva Show, and duly took the *Carrosseries Speciales* award at the event. As Bertone's press release stated: "Bertone's styling in this case is mostly concerned with elegance, sobriety and overall effectiveness, without straying into anything which

Bertone announced the Ascot prototype at the 1977 Geneva Show. Compared with the standard XJ-S, the Ascot was shorter, wider and lower, and had the straight-edge styling that was very much en vogue at the time.

might be of impact to the onlooker. The result is a refined harmony, well-balanced throughout ..."

According to Bertone's own paperwork on the model, the 12-cylinder XJ-S was chosen as the basis for the Ascot to add prestige to the vehicle. "Prototypes either derived from small cars or built around 'ordinary' mechanics are not sufficiently appealing to the imagination of the audience," it said.

Most people seemed to embrace the design, and were particularly taken by the soft leather- and suede-trimmed interior features. The Ascot was 230mm (9in) shorter than the standard XJ-S, but 80mm (3.1in) wider and 60mm

Much of the standard instrumentation was retained in the Ascot, although the removable radio was a novel feature at the time, as was the mobile telephone.

The stunning XJ-S Spyder by Pininfarina.

Interior of the XJ-S Spyder.

Tail-end treatment of the Pininfarina show car. Jaguar's own XJ41/42 would look very similar, but that, too, was destined never to go into production.

(2.4in) lower. Although this was not Bertone's best work (despite being from a different era, there is little that can compete with the sheer drama and elegance of Bertone's Lamborghini Miura), it was certainly a stunning piece of the coachbuilder's art.

For the 1973 Paris Salon, Pininfarina produced an elegant four-door saloon based on the XJ12, but at the 1978 Motor Show - the first to be held at the NEC in Birmingham - Pininfarina unveiled something far more spectacular - the XJ-S Spyder. Based on an ex-development car provided by Browns Lane, it bore a distinct resemblance to the aerodynamic study presented at the last Turin Show, and was in complete contrast to Bertone's Ascot, which followed what Bill Mitchell, GM's top stylist, called the "folded paper" school of design.

Indeed, designs featuring straight edges were becoming more and more popular as the 1970s progressed.

Styled by Leonardo Fioravanti, Sergio Pininfarina and Renzo Carli, it was the star of the show. One magazine noted: "Many were heard to remark that it is the sort of car the XJ-S

should have been: an up-to-date E-type. Certainly it was quite deliberately styled to bear a family resemblance to that most illustrious of sports cars and to its forebear, the D-type."

The interior and fascia were very futuristic, but unlike so many show cars, there was actually a small chance

73

In the early part of 1979, the manual gearbox option was officially struck from the lists. This particular vehicle dates from that period, although even now, the styling was not universally accepted. Tony Rudd of Lotus fame once said: "It looks as if three people designed it and they weren't talking to each other!"

that the Spyder could have entered production. However, with BL strapped for cash, sadly, it was never to be.

The Series 3 Saloon

In early-1979, the XJ-S manual gearbox option was officially deleted from the price list. It had never been popular, and only 352 cars had left the factory with the four-speed transmission at this time, although another four manuals were produced during 1980 to take the total to 356. A new Lucas-Bosch digital fuel-injection system was introduced towards the end of the year,

giving a slight improvement in fuel economy and the engine's maximum torque output.

America's *Motor Trend* waxed lyrical over the V12 power unit: "The XJ-S also follows Jaguar's performance tradition. With the exception of the 928, it is the class of its field in this area. The 244-horsepower V12 responds instantly and impressively throughout its full range, with no hesitation or flat spots. Though the car registered superior quarter-mile times, its most satisfying charge comes in the mid-range speeds. A quick jab of the

throttle and you're around slower moving obstacles in a flash.

"The flaw in the Jag's performance is its braking. Even with four-wheel disc brakes, its nearly 4000lb bulk is reluctant to come to a smooth, rapid stop under the most extreme conditions. In normal driving, braking was adequate and presented no problems. But during track tests, where we try to bring the car to the quickest possible halt, we encountered a lack of stability and the need for very deft brake pedal manipulation."

Meanwhile, in March 1979, the

Pininfarina gave the XJ series subtly different lines for March 1979. The basic shape remained, but the Series 3 cars had a crisper, more modern look.

Series 3 XJ Jaguar-Daimler saloons were launched on the English Riviera. The Series 3 came with three engine options for the Jaguar range (3.4, 4.2 and 5.3 litres), whilst the Daimlers were listed with only the two larger units. Manual five-speed or three-speed automatic gearboxes were available on the six-cylinder cars, but the V12 came with automatic transmission only. As well as subtle styling changes (executed by Pininfarina of Italy), the interiors were upgraded, too, with a whole array of new features and equipment.

However, even in the mid-1970s, work on their replacement was already well in hand. Because Leyland (or BL Cars, as it was now known) wanted the Rover V8 to be fitted rather than the thirsty V12, the Jaguar engineers deliberately made the inner front wing area on XJ40 too narrow to accept a V-engine. This last ditch attempt at self-preservation later backfired, but it shows the incredible loyalty that certain members of the workforce had towards the marque.

As usual, there was bleak news from the BL camp: 13 plants were to close, including the MG factory at Abingdon. Ironically, given that the British industry had lost so much of its market share to the Japanese, BL signed an agreement with Honda to build the Ballade in the UK (badged as the Triumph Acclaim), and later on, the Legend would form the basis for the Rover 800 executive saloon. These were indeed strange times ...

"The fastest production car in America"

So ran the headline in the February 1980 issue of *Car & Driver*. Fastest car on the road, maybe, but the S was very slow from a sales point of view. Just over 1000 cars were sold in the States in 1978, but this figure dropped to 695 the following year, before dropping again to 420 in 1980. Now costing $30,000 in automatic form, the coupé's styling was "beginning to show its age, though it still snares its share of stares, and the inside is a place of leather-lined comfort and convenience."

Performance, however, was never in doubt. As *Car & Driver* noted: "Rated horsepower for the 1980 XJ-S is up from 244 at 5250rpm to 262hp at 5000rpm. The V12's compression ratio has been pressured up to 9.0:1 from 7.8:1, and the wonders of modern science are running rampant all around the engine in the form of Lucas-Bosch electronically-controlled (D-Jetronic) fuel-injection, which uses closed-loop feedback and a three-way catalyst to fry emissions."

With the rear axle ratio having changed from the original 3.31:1 to 3.07:1, in one test, the Jaguar coupé recorded a top speed of 143mph, with 0-60 coming up in 7.9 seconds. Fuel consumption was still less than ideal, the EPA rating being just 13mpg, but as Larry Griffin observed: "The XJ-S is for getting down to business ... And, in doing so, it is of no small importance that the XJ-S will make the rate of your trips and the rate of your heart ever so much faster."

Standard colours at this time included Tudor White, Cotswold Yellow, Damson Red, Racing Green Metallic, Silver Frost Metallic (which replaced the recently-introduced Rhodium Silver Metallic), Cobalt Blue Metallic,

Quartz Blue Metallic, Chestnut Brown Metallic and Sebring Red.

A convertible

The first company to make a soft-top version of the S was an American firm, Royal Carriage Motors. In early-1980, it announced the XJ Phaeton (or XJ Phaeton Par Excellence in luxury trim), limited to just 250 units. An interesting customer service provided by the company was a five-speed ZF manual gearbox conversion but, this being the States, most buyers stuck with the standard automatic transmission.

Hot on the heels of the Washington State concern was England's Lynx Engineering, with a highly professional conversion that definitely caught the attention. Lynx (and Avon Coachworks) had produced convertibles on the Series 2 two-door models, so were well-qualified to carry out the task.

As *What Car?* said in 1980: "The prototype XJS Spyder is the result of nearly two years' development work, for as everyone knows the removal of a unitary construction car's roof deprives the vehicle of much of its rigidity. Additional strengthening panels are welded in the sill and door pillar areas, as well as behind the rear seat, though despite the presence of a bulky, power-operated hood, rear-seat shoulder width is only fractionally reduced and both headroom and legroom (what there ever was of it, at least) are unaffected."

The Lynx conversion incorporated electric rear quarter windows which, when lowered, gave the convertible beautifully clean lines, as the fully-lined mohair hood sat almost flush with the car. In its test of the vehicle, *What Car?* noted: "The Spyder gave remarkably little turbulence or buffetting. Some scuttle shudder occurs - this is sensed rather than felt - but some loss in rigidity has clearly come about somewhere along the line." However, designed by Guy Black (an ex-Weslake engineer), it was an impressive conversion; taking around ten weeks to complete, the price was set at £6950 plus VAT.

Interestingly, that other BL product whose design was so influenced by the proposed ban on the convertible, the Triumph TR7, was now available from the factory as a soft-top. One wondered how long it would be before an official S drophead was launched.

Egan's appointment

Asked about his views of the Leyland days, Jim Randle (who worked on the XJ-S chassis and was the man behind XJ40) told the author: "Up to the days of Michael Edwardes it was terribly wasteful. The industry had to get together to survive, but it was all being driven by a political agenda - done by people who knew nothing about the industry. The opportunity was lost, as I believe the mergers took place ten years too late. The Ryder Report, which everyone knew was wrong, was carried through regardless. Edwardes did a remarkable job breaking the industry back down, but it was too late and nothing really changed."

By 1980, it had become all too evident that the Jaguar-Rover-Triumph arrangement wasn't going to work either. Quality of workmanship was very poor, and, in fact, became so embarrassing that the company stopped reporting to the SMMT the number of faults per car found on inspection. Morale was at an all-time low, and the strike that followed seemed fairly inevitable. It was brought to an end only when Sir Michael Edwardes told the workers to return or risk losing their jobs.

Being Jaguar's Chief Executive must have seemed like a nightmare at the time, but John Egan (who had created Unipart) decided he would rise to the challenge. As Jaguar's first full-time head for over five years, it must have been a daunting prospect. On the second time of asking, Egan accepted the job in March 1980. He returned from a skiing holiday to take up his new post at Browns Lane on the first day of April, right in the middle of a strike! Having made the bold statement that "one cannot have better ground to build on," he set about getting the Shop Stewards and union people to fight on the same side. Within a very short space of time, it became obvious that this was the man who would make Jaguar great again.

The future was still uncertain, though, especially for the XJ-S. The XJ-S was never meant to be a replacement for the E-type, although the fact was never made clear by the Leyland officials that this was the case. Perhaps wanting the GT to benefit from the assumption that most people made - that it was the new E-type - from initial reactions, the plan backfired, as the S was so different that, if judged by sports car criteria, it would always

These three photographs were released in 1980 to try and determine which image appealed most to the prospective XJ-S owner: "glamour and romance, sleek styling in solitude or high-speed action." The results were to form the basis of a promotional campaign to give the car what was, in effect, one last chance at survival.

come second best to the legendary E, and with another oil crisis about to bite, the thirsty V12 engine was less than ideal, to say the least. As *Motor* said in the 25 October 1980 issue: "If 'efficiency' is the watchword for the 1980s, what hope is there for the Jaguar XJ-S?"

In reality, most thought the Grand Tourer was doomed from the start. One member of the Jaguar staff described the car as "a white elephant" and even Sir John Egan went on record saying: "I'd had a Series Three [saloon] and knew it could be made into a saleable car, though I wasn't so sure about the XJ-S."

JAGUAR
XJ-S

4

A CALL FOR
EFFICIENCY

> *"Our job, our mission, is simply to build magnificent motorcars, cars of the highest possible quality in every way."* **John Egan, Chairman, Jaguar Cars.**

John Egan was born in 1939 and educated at Bablake School in Coventry. He moved to Browns Lane from Massey Ferguson (the tractor manufacturer), so wasn't a dyed-in-the-wool Jaguar man, but after Sir William Lyons, he was probably the most influential leader the company has ever had. He took a no-nonsense approach to both Jaguar's products and working practices, somehow managing to retain tradition whilst moving the ailing concern forward.

It was a frightening situation, and one that has never been confirmed, but Jaguar was extremely close to being shut down. Every time that Leyland ran short of money and was refused extra funding from Mrs Thatcher's government, it closed down a site; Speke went, the Kingfield Road Service Department went, then Canley as a production plant, and eventually the SD1 line moved from Solihull to Cowley. With Longbridge and Cowley being too politically sensitive to close, Jaguar was next in line. At a meeting of the Board, Egan was the only man there who stood up for the Coventry company: he was given one month to jus-

Sir John Egan, the man who turned around the fortunes of Jaguar. From a £47 million loss in 1980 when he arrived at Browns Lane, healthy profits were recorded on the balance sheets within three years.

tify keeping it going, which he did against all the odds.

However, Egan got to the root of the problem by attacking quality control and bringing back customer service at both the works and within dealerships. If the cars had troublesome areas, he said 'deal with it' - let it be a Jaguar problem, not a customer problem. Certain German manufacturers have followed this philosophy for years, and it gives the end user the feeling of ultimate reliability, even if that is quite often not the case.

Egan fought Leyland and won, creating at Browns Lane a feeling of pride that had been lacking for many years. Supplier problems were sorted (over 60% of reported failures were due to bought-in parts), and the former Pressed Steel Fisher body plant at Castle Bromwich was put under Jaguar's control following the end of the Triumph TR7. This was another important move in the fight to improve quality. As Mike Cook of Jaguar Cars Inc. in America once said: "What they accomplished [at Jaguar] was marvellously effective."

But what of the XJ-S? Egan knew as well as anyone that, if the XJ-S was to survive, it needed some radical changes. During 1980, production was stopped to try and reduce the stockpile (1760 cars were sold in that year, compared with a production figure of just 1057), and there were even thoughts of fitting the six-cylinder XK engine to increase sales volume.

The 1981 model year

For 1981, Jaguar introduced a number of running changes on all V12-powered models. The most important was the use of a microprocessor to control the P-Jetronic fuel-injection system, which not only provided better economy, thanks to more accurate fuel metering, but had the added bonus of cutting exhaust emissions. In turn, this allowed the compression ratio to be raised from 9.0:1 to 10.0:1, leading to a 5.4% boost in power output and an 8.2% increase in maximum torque.

The V12 now produced at 300bhp at 5400rpm and 318lbft of torque at 3900rpm. Naturally, performance on the road was enhanced, and tests on the XJ-S (named 'The Silent Sports Car' by *Classic Cars* - a slogan used by Bentley to describe its sporting models of the 1930s) illustrated that the gap between the later automatic and original manual cars on the press fleet was much narrower than previously: the 0-60 time was just 7.6 seconds, whilst the standing-quarter was covered in 15.8; top speed was in the region of 145mph.

It wasn't all about brute force, though. As *Motor* commented: "What no figures can convey, however, is the quality of the transport the XJ-S provides. One of our testers commented that its V12 engine must surely be the quietest and smoothest production engine ever made, and it's hard to think of one that even compares. Equally remarkable is the sheer breadth of its powerband and total lack of temperament. All qualities which combine to create an experience you only ever encounter in a V12 Jaguar: super-quiet, silky and effortless potency. It's addictive, be warned."

To enable the driver to keep performance in check, the XJ-S was endowed with "progressive and tremendously powerful" brakes, "capable of hauling the heavy XJ-S to rest from high speed without drama or fade, wet or dry." It was good to see there was no mention of brake judder, or other problems some testers had noted in earlier days.

The staff at *Motor* still thought the steering was a little on the light side, but "its crisp responses and sensible gearing do help compensate, creating a sense of wieldiness that belies the car's size and weight."

As for the interior, they were not overly happy with the rather "bland fascia" and the thin silver line sur-

rounding the gauges, stating it "made an otherwise attractive display look cheap." There was also concern regarding the air conditioning unit, which was "poor in its inability to provide proper bi-level heating/ventilation." At least the front seats were praised - although they looked thinly-padded, they were "actually well-shaped and comfortable," something anyone who has ridden over long distances in an XJ-S will verify.

Priced at £19,187 in October 1980, perhaps its nearest rival in the UK was the BMW 635 CSi. The Bavarian coupé, the first of the breed being introduced at the 1976 Geneva Show, had similar

performance and, with this in mind, excellent fuel economy; it was also a couple of hundred pounds cheaper. Most other potential rivals were either far more expensive or simply couldn't match the XJ-S's performance.

Bill Boddy of *MotorSport* suggested: "There would seem to be no good reason at all for any Britisher who is in the market place for a high-performance car to rush off and buy a foreigner; unless he is seeking a raspy, rorty sort of GT machine - the Jaguar's V12 power unit is muted indeed, by comparison. The Jaguar XJ-S is certainly a motor car that I would gladly own ... after driving one I find it aston-

ishing that there are not more on the road."

In America, the XJ-S was the only V12 car officially available on the US market - the Ferrari 400i wasn't certified for US sale and the Lamborghini Countach wasn't sold there from 1977 to 1982, although a few did make it to the States as 'grey imports'. Even the V12 Jaguar saloons had been dropped from the American market when the Series 3 models arrived, so there was a real kudos attached to the S.

Like the European specification engines, the Federal V12 was also given more power. As *Road & Track* noted in May 1981: "This latest version replaces a pair of oxidizing converters, air injection and exhaust gas recirculation with a single three-way converter linked via oxygen sensing to the Lucas/Bosch digital electronic fuel-injection. Because of these changes in emission control, and an increased compression ratio (7.8 to 9.0:1), horsepower increases from 244bhp at 5250rpm to 262 at 5000; torque jumps from 269lbft at 4500rpm to 289 at 4000. And the engine positively purrs under almost any condition."

In American trim, the $30,000 machine would cover 0-60 in 7.8 seconds and go on to pass the quarter-mile marker in 15.9 seconds (both slightly better than the times posted by the Porsche 928). With the 3.07:1 final drive, a maximum speed of 63mph was possible in first gear, 104 in second and 139 in top.

It is interesting to note that, officially, there wasn't a 1981 model year XJ-S in America. Cars sold in the early part of '81 were classed as late 1980 models, as an upgraded 1982 version was imminent, scheduled for a mid-year launch. As a result, just 232 cars were sold in the States in 1981! Thankfully, in July 1981, a new, more efficient power unit was announced.

A new engine

The V12 HE (High Efficiency) engine was introduced on 15 July 1981. With tighter worldwide emission controls, and the need to improve fuel economy, it was a brave decision to invest in the V12, but the 299bhp HE engine replaced the old 12-cylinder unit across the whole Jaguar-Daimler range.

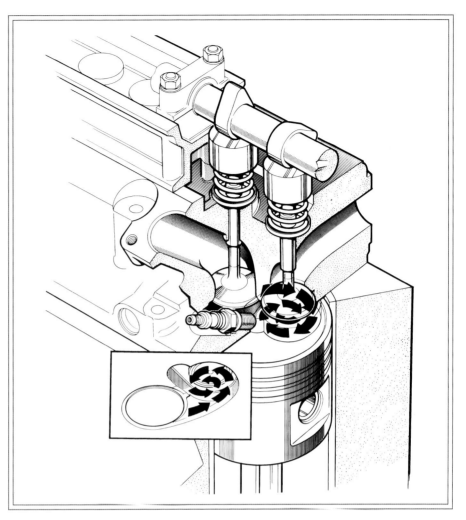

The key to the HE engine was the Fireball cylinder head. Due to the combustion chamber design, the V12 unit now gave a much more efficient and cleaner burn.

The main difference was the Fireball cylinder head design, an adaptation of a patent filed by the Swiss engineer, Micheal May. May had adapted a Volkswagen Passat engine to demonstrate his theory, and, in the early part of 1976, Jaguar's Harry Mundy travelled to Switzerland to observe the results May was claiming.

Mundy was very impressed: it was just what the Coventry firm needed to revive the V12, and an agreement was reached enabling Jaguar to use May's patents. A single-cylinder prototype was produced, followed by a 2.6 litre slant-six that was basically half of the V12 unit to help refine combustion chamber and piston shapes. Naturally,

there were a number of teething troubles, meaning work on a V12 unit didn't take place until 1979, but the end result was remarkable. With a combination of high turbulence in the split level combustion chamber, a very high compression ratio (Mundy had the V12 running happily on a 14:1 c/r on the testbed) and remarkably lean fuel/air mixture, it translated into lower emissions and a healthy improvement in fuel consumption - as much as 20% at some engine speeds.

The V12, with its 12.5:1 compression ratio in European trim, produced 299bhp at 5500rpm and 318lbft at 3000rpm. Although these figures weren't that different to 1981 specifi-

cations, with maximum torque peaking much lower down the rev range, the engine felt more powerful than the old unit, and allowed a taller rear axle ratio to be used (2.88:1 instead of 3.07).

With such a transformation under the bonnet, Jaguar took the opportunity to refurbish the rest of the XJ-S as well. The V12 coupé was still only available with the three-speed GM400 gearbox, but now, thanks to some revalving, first gear could be selected manually or obtained via kickdown whilst the car was moving. Contrary to what it says in the workshop manual,

the early GM box wouldn't engage first gear at speed.

The HE was distinguished from the outside by the new 6.5J 'Starfish' alloy wheels (slightly wider five-spoke items fitted with 215/70 VR15 Dunlop tyres), chrome trim to the upper surface of the restyled bumpers, a double coachline running along the waistline of the vehicle, and new badging that included a round Jaguar 'Growler' on the bonnet.

The interior was also upgraded to a higher standard, more in keeping with the Jaguar marque. At the request of American dealers, more leather

was added to the console, rear side and door trims, whilst lighter-coloured burr elm veneer (rather than the traditional walnut) was used on the restyled fascia panel and door fillets. Along with a Philips radio/cassette, air conditioning was still a standard item. Amazingly, the price was £18,950 on introduction, some £800 cheaper than the outgoing model.

Autocar reported on the HE in April 1982, by which time the price had increased somewhat. Although Jaguar had claimed a top speed of 155mph, the testers couldn't better 151. However, a later test from the same journal gave a best of 157mph, so the factory was fully justified in stating that the XJ-S had supercar credentials. Even at 151mph, it was still declared the world's fastest production car with an automatic transmission, a title reinforced by a sub-15 second standing-quarter and 0-60 time of just 6.5 seconds.

Comparing the £19,708 HE with cars like the Aston Martin V8, BMW 635 CSi, Ferrari 400i, Maserati Kyalami and Porsche 928S, *Autocar* said: "For all-round satisfaction - a tiger of a car when need be, the most refined, relatively speaking, of Grand Touring 2+2s when driven less hard, let alone gently, and always a gentleman in character - the Jaguar is even more so the paragon, especially now that its fuel economy is so much better than before. It is also more motor car for the money than the others, in every way."

It was not only the quickest of the models looked at, but also one of the most economical - only the Porsche

One of the earliest HE models. Note the new alloy wheels and bumpers.

and BMW were better overall when it came to fuel consumption, although the BMW couldn't compete in the performance stakes. Add in the fact that the XJ-S was now beautifully finished, yet half the price of the Aston, and substantially cheaper than all the others except the BMW, and one can see how the magazine reached this conclusion.

It wasn't all good news, however. In a comparison test between the new Jaguar, the £25,250 Porsche 928S and £28,700 Mercedes-Benz 500SEC, *Motor* found the German cars superior in most respects (the Jag had 59 out of 75 points, compared to 64 for the Porsche and 66 for the Benz), but being more than £5500 cheaper than its nearest rival, *Motor* found the Jaguar "a lot of car for the money."

However, John Bolster of *Autosport* was delighted with this latest product from Browns Lane: "Perhaps the most delightful feature of this Jaguar is the wonderful reserve of power. One seldom uses more than half throttle and it is all too easy to wander along at 120mph when 70mph was intended, unless cruise control is employed. I have previously praised the silence, smoothness, and refinement of the car, allied with its extraordinary performance and riding com-

fort. Perhaps no other car is less tiring when driven on a long journey."

With improved quality (*Motor* awarded full marks for the test car's finish - something that would have been unheard of five years earlier) and economy, sales of the V12 coupé started to rise steadily, from 1199 in 1981 (an all-time low for a full year) to 3111 in the following 12 months. More importantly, sales continued to rise, thus ensuring not only the future of the legendary V12 power unit, but the future of the XJ-S as well.

American revival

"Jaguar picked a great way to celebrate its 50th anniversary. The 1982 XJ-S is a luxury GT that sports the most powerful, efficient, and sophisticated V12 engine Jaguar has ever built. Its wrappings are of the highest quality, and it is the most completely packaged S since the car's introduction in 1975," said *Motor Trend* .

In American trim, although the twin coil ignition and Lucas/Bosch digital fuel-injection system on the European cars was retained, an 11.5:1 compression ratio and battery of catalytic converters reduced the HE engine's power to 262bhp at 5000rpm and 290lbft of torque - still the highest output of any car sold in the US then.

Performance was excellent - 0-60 in a fraction over eight seconds, with the standing-quarter being covered in 16.3 seconds. Fuel consumption was said to have improved by an average of 10%. More importantly, given that petrol is so cheap in America anyway, with its new-found efficiency, the XJ-S managed to avoid the US gas-guzzler tax, which would obviously have a significant bearing on sales.

From the outside, US specification cars looked much the same as their European counterparts, with the new bumpers, badges and wheels. The only real differences were the front side markers, a lack of the heavy double coachlines applied to models for other markets, the now-familiar Federal headlights and Pirelli P5 rubber instead of Dunlop D7 tyres.

With a price tag of $32,000, one would expect something special from the interior. In addition to extra leather and wood, standard equipment for 1982 included a high quality, four-speaker stereo radio/cassette, interior courtesy lights with delay, door-edge warning lights, electrically-adjustable mirrors on both doors, intermittent wipers and better carpet in the boot. The manufacturer's statement "Optional Equipment Available: None" confirmed the luxury specification, and few were disappointed. *Motor Trend* was captivated by the improvements inside the car: "The first thing that struck our fancy about the new XJ-S was its new interior, which, with all that burr elm and additional leather, looked more traditionally British than ever. The seats are a compromise be-

Advertising introducing the HE model to the American market. The interior was definitely more in keeping with Jaguar tradition than that of earlier cars.

tween all-out luxury and sportiness, but can't be considered 'driving seats' in the new argot. They move fore and aft and the backs recline, and that's it. No adjustable bolsters, no lumbar adjustment, no tilt feature. Yet somehow the seats are supremely comfortable without all the knobs, handles and cranks. The elm trim adds luxury and warmth to an already sumptuous interior, and was fitted and finished beautifully."

Summing up its review of the new model, *Road & Track* stated: "The Jaguar XJ-S HE is a car that can lay legitimate claim to full credentials as one of the world's very finest 2+2 Grand Touring automobiles. It has the requisite supple but controlled ride, the superb handling and braking, and the ability to instill that extra feeling of confidence and safety. It also has that wonderfully smooth flow of power, and a slightly heavy feel to the whole machine that gradually lightens up as speed increases. It is, without question, an excellent example of the breed."

Egan's drive for quality was making an impression. Speaking in 1981, he said: "We identified 210 significant faults from warranty claims. Resolving 150 of them gave us the same quality and dependability as Mercedes and BMW." Quality became a key word once again at Browns Lane, and, instead of dealing with warranty work through Cowley (which had a limited budget to deal with the whole British Leyland range), it was brought back to Coventry and strictly monitored; investments were made in robotics, further improving build quality.

A US advert from the end of 1982. American sales had slowed almost to a standstill before the introduction of the HE, but the S (and the Jaguar marque as a whole) was to stage a remarkable comeback. Better quality was the key to Jaguar's success following John Egan's appointment.

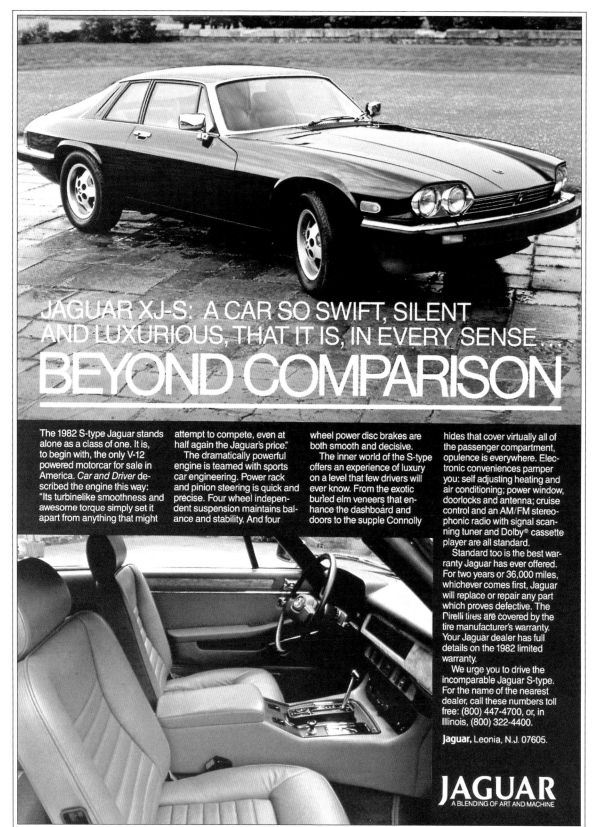

JAGUAR XJ-S: A CAR SO SWIFT, SILENT AND LUXURIOUS, THAT IT IS, IN EVERY SENSE...

BEYOND COMPARISON

The 1982 S-type Jaguar stands alone as a class of one. It is, to begin with, the only V-12 powered motorcar for sale in America. *Car and Driver* described the engine this way: "Its turbinelike smoothness and awesome torque simply set it apart from anything that might attempt to compete, even at half again the Jaguar's price."

The dramatically powerful engine is teamed with sports car engineering. Power rack and pinion steering is quick and precise. Four wheel independent suspension maintains balance and stability. And four wheel power disc brakes are both smooth and decisive.

The inner world of the S-type offers an experience of luxury on a level that few drivers will ever know. From the exotic burled elm veneers that enhance the dashboard and doors to the supple Connolly hides that cover virtually all of the passenger compartment, opulence is everywhere. Electronic conveniences pamper you: self adjusting heating and air conditioning; power window, doorlocks and antenna; cruise control and an AM/FM stereophonic radio with signal scanning tuner and Dolby® cassette player are all standard.

Standard too is the best warranty Jaguar has ever offered. For two years or 36,000 miles, whichever comes first, Jaguar will replace or repair any part which proves defective. The Pirelli tires are covered by the tire manufacturer's warranty. Your Jaguar dealer has full details on the 1982 limited warranty.

We urge you to drive the incomparable Jaguar S-type. For the name of the nearest dealer, call these numbers toll free: (800) 447-4700, or, in Illinois, (800) 322-4400.

Jaguar, Leonia, N.J. 07605.

JAGUAR
A BLENDING OF ART AND MACHINE

Bottom left: Earlier legislation in Germany cast a shadow of doubt over sales of the XJ-S in Germany (the car's infamous flying buttresses were thought dangerous), but this was later resolved. This is a German-registered HE model.

The test we quote in the headline also described the XJS-H.E. as "travelling so quickly and silently between two points that some other means than internal combustion seems to have been employed." Which you may think says it all.

However, we would like to add a fact or two.

First, the engine. 5.3 litres, V12, 300 BHP with a maximum speed of 155 mph, and acclaimed as the finest production engine anywhere in the world.

Yet the XJS-H.E. delivers fuel economy comparable to many smaller engined cars. (On a run from Munich to Calais Motor Magazine's test car achieved 21.9 mpg at an average running speed of 62.6 mph.)*

This remarkable achievement is made possible by the adoption of the unique 'fireball' cylinder head, combined with an advanced digital fuel injection system which precisely

meters the amount of fuel required second by second.

The suspension system is so smooth and the sound insulation so complete that you can carry on a civilised conversation at speeds almost embarrassingly higher than the legal limit.

And naturally, in the best Jaguar tradition, we haven't forgotten creature comforts.

Connolly leather seating and trim, burr elm veneer dashboard and door cappings, deep pile carpeting, air-conditioning, and a four-speaker stereo/radio cassette system, together with many other features combine to give an unequalled level of interior refinement.

The overwhelming verdict of the motoring press is that the XJS-H.E. is the finest Grand Touring car money can buy.

We'd like to add that the few rivals for that title can cost up to twice as much as the Jaguar's price of £20,693.

"THE FASTEST AUTOMATIC TRANSMISSION PRODUCTION CAR IN THE WORLD."
SUNDAY TIMES, OCTOBER 1981.

SUPERCAT. THE XJS-H.E.

Price, based upon manufacturer's RRP and correct at time of going to press, includes seat belts, car tax and VAT. (Delivery road tax and number plates extra). Official fuel consumption figures: constant 56 mph 27.1 mpg (10.4 L/100 km); constant 75 mph 22.5 mpg (12.6 L/100 km); urban cycle 15.6 mpg (18.1 L/100 km); *issue dated 18th July 1981.

JAGUAR
In pursuit of perfection.

One of the better British adverts from the early 1980s, this one dating from May 1983. Note that the 'Jaguar' script on the bootlid was now silver on black, rather than black on silver as on the earlier pre-HE cars.

A 24 month/36,000 mile warranty was established, and as a result, with renewed confidence in the marque, sales reached record heights in America: in 1979 and 1980, less than 3000 cars were sold to the States but, by 1982, helped by 1409 XJ-S sales, the company had broken the 10,000 units per annum barrier for the first time in Jaguar history.

In 1983, despite a $2000 hike in price, XJ-S sales in the US almost doubled. In fact, the 2705 models sold in that year represented a higher figure than total sales for 1975, 1977, 1979, 1980 and 1981! Jaguar was definitely back on track.

The six-cylinder XJ-S

The new AJ6 (Advanced Jaguar Six) engine was announced in autumn 1983, and formed the basis for a whole series of new power units. With a capacity of 3590cc (91 x 92mm), the

new six started life in 1974, but went through a number of changes and a £30 million investment to become the engine we know today. Developed under the watchful eye of Harry Mundy, and then, from 1980, Trevor Crisp, the AJ6 employed four valves per cylinder, operated by twin-overhead camshafts. Electronically-controlled Lucas fuel-injection and a 9.6:1 compression ratio helped the engine to develop a healthy 225bhp at 5300rpm, with maximum torque put at 240lbft.

In October 1983, the 3.6 litre XJ-S coupé (code XJ57) made its public debut, the first car to feature the AJ6. As the first all-new six-cylinder Jaguar engine since the XK made its debut in 1948, it was only right it should have impressed the *Autocar*: "It is the 24-valve engine's flexibilty that impresses most. Such is its low-down pulling power that, in town and urban conditions, one rarely needs more than 2000rpm to keep up with the traffic flow."

However, a number of testers - including the author - felt the new AJ6 was a little harsh compared to the old XK lump. Of course, the old 4.2 litre engine was developing a rather lazy 205bhp at the time, but this lack of stress on the unit made it very refined. It should be said, though, that the vast majority of journals praised the AJ6, which was canted over some 15 degrees under the XJ-S bonnet.

With a manual five-speed transmission from the Getrag concern and a 3.54:1 final drive, this latest S was capable of covering 0-60 in 7.2 seconds before going on to a top speed of

The 3.6 litre version of the AJ6 (Advanced Jaguar 6) engine, seen here in cutaway form.

This photograph shows the AJ6 unit installed in the XJ-S engine bay. (Courtesy Paul Skilleter)

A 3.6 litre coupé dating from about a year after the launch. There was little to distinguish the six-cylinder model from the V12 except the alloy wheels and badging. Those with sharp eyesight will spot that the grille badge is different to that of the larger-engined model.

Interior of the new 3.6 litre coupé. Note the manual gearbox. (Courtesy Paul Skilleter)

One of the Cabriolet bodies being checked for accuracy on a marking off machine at Coventry's Park Sheet Metal. Founded in 1947, PSM was also building the Daimler limousine body for Jaguar at the time, along with the shell for the Rolls-Royce Camargue.

137mph; factory figures stated the six-cylinder car should be capable of 142mph, which wasn't that far behind the V12. At the same time, fuel economy was significantly better than with the V12 model: 36mpg at a steady 56mph being quoted.

There was little to identify the 3.6 from the outside except the badges on the bootlid and the new 6J 'Pepperpot' alloys. It still had many of the luxury features carried in the V12, such as air conditioning, remote control door mirrors, electric windows, central locking, and so on, but the trim was now predominantly ambla with just the seat facings trimmed in leather.

The 3.6 had softer springs and a slightly thinner anti-roll bar up front (the AJ6 was naturally much lighter than the V12 engine), and no anti-roll bar at all at the rear. There was a fractionally stiffer torsion bar in the steering with the objective of giving a more sporty feel, for as Paul Walker, one of the leading engineers on the project, said: "We wanted a more responsive car that turns in a little faster for the sporting driver."

Basically, they succeeded. "Although more realistically weighted than it used to be," said *Motor*, "XJ-S steering remains largely devoid of feel, effort at the steering wheel rim staying constant even when the front wheels are beginning to lose adhesion on a slippery surface. This is at odds with the steering's crisp responses and sensible gearing, which help create a sense of wieldiness that belies the car's size and weight."

At the end of 1983, the manual 3.6

litre coupé cost £19,248, which was around £2500 cheaper than the £21,752 XJ-S HE. Options included a headlamp wash/wipe system, a stereo radio/cassette and a trip computer, all of which were standard on the V12; the buyer could also specify rear seat belts, and black or metallic paint was a no-cost option on the HE model.

A Cabriolet

At the same time as the 3.6 XJ-S coupé was announced, Jaguar introduced the XJ-SC - a 3.6 litre cabriolet model. After Abbey Panels had built five prototypes to establish which design would give the best compromise on strength versus open car styling, it went into production with twin removable targa panels for the front part of the roof and, for a while at least, a removable glassfibre hardtop for the rear, which could then be covered by the half-length cloth hood.

The body was produced at Castle Bromwich as usual, but the conversion work was carried out by Park Sheet Metal in Coventry (where the Daimler limousine shell was made). After the roof and flying buttresses had been removed, the body was suitably modified to make it into a drophead. The shell was then painted at Castle

Bromwich before the car joined the line to have its components fitted. Eventually, the targa panels and hood were fitted a few miles up the road at the recently-opened Aston Martin Tickford factory in Bedworth, and a final inspection took place at Browns Lane; it was a very long-winded affair.

Built to order, the Cabriolet (known internally as XJ58) was a pure two-seater, as Jaguar was fearful of a claim under the USA's product liability laws should rear seat passengers be hurt on the roof structure in the event of an accident. In any case, the hood and roll-over bar deemed necessary to retain the vehicle's strength, made it difficult to fit occasional rear seats, so lockers were built in behind the front seats instead, although TWR did offer a rear seat conversion for £1389 plus VAT. Many of the V12 features were carried over onto the Cabriolet, such as the 'Starfish' alloys, full leather trim (six hides were used in all), and so on.

Again, this 3.6 litre model was available in manual form only. *MotorSport* noted: "The Getrag gearbox never demonstrated any reluctance or stiffness whatsoever and, in our view, a manual transmission undoubtedly enhances the XJ-S range's overall appeal."

The XJ-SC 3.6 Cabriolet as it first appeared in October 1983. The rear section could be lowered to give the next best thing to a full convertible.

Another view of the same model, this time with the roof panels in place. Some have said that the XJ-S looked better without the controversial 'flying buttresses'; for what it's worth, the author still prefers the coupé.

Being only a few kilos heavier than the 3.6 coupé, performance was very similar. However, as *Drive* pointed out: "Those expecting a gut-wrenching supercar performance machine are going to be disappointed by the XJ-S Cabriolet. A gentleman's high-speed carriage is the description we heard most often from drivers who appreciate the finer things of life."

At £20,756, it offered excellent value for money, but the Cabriolet was, of course, something of a stop-gap, as a full convertible was always going to be a more popular form of

Interior of the 3.6 litre Cabriolet. Note the traditional XJ-S handbrake location to the right of the driver on this UK specification model. (Courtesy Paul Skilleter)

way with all the panels removed - it looked bitty. However, like the AJ6-powered coupé, it was well-received by press and public alike.

At the end of its time with the Cabriolet, *MotorSport* stated: "It has been our pleasure to sample what must be one of the most attractive machines in the range, the XJ-SC 3.6 - basically an XJ-S cabriolet fitted with the latest AJ6 six-cylinder engine and mated to a delightful five-speed Getrag manual box. The result is a magnificent blend of almost regal, boulevard splendour and sports car agility: arguably, an all-round package which is even more appealing than the out-and-out brute power of the V12-engined XJ-S."

Motor noted: "The timely arrival of two new and more frugal XJ-S models - one a convertible and both powered by a new generation, all-alloy, 3.6 litre straight-six - not only opens up a fresh chapter for Coventry's Big Cat but also gives Jaguar the ammunition to make a deeper dent in the lucrative luxury car market. Less powerful [than the V12] but lighter, and equipped with five-speed manual transmission, the six-cylinder cars, though not that much cheaper, step more assertively into BMW/Mercedes/Audi territory."

Although the V12 continued to be the best seller in the XJ-S line-up (it should be remembered that, without the new-found sales success of the HE coupé, these 3.6 litre models would not have been possible), the six-cylinder models were nonetheless increasing sales and giving some useful field testing of the AJ6 unit. Less than 250 six-cylinder XJ-Ss had been built by the

A final shot of the 3.6 litre Cabriolet. However, introduction of the 3.6 litre models weren't the only XJ-S changes for the 1984 model year, as the HE model received a headlamp wash/wipe facility, a trip computer, digital stereo radio/cassette and cruise control as standard.

open-top motoring. From a styling point of view, most people agreed that the Cabriolet looked better than the coupé.

The author is not so sure, for while it looks very pretty with the hood up, there were just too many bars in the

American advertising for the 1984 S-type.

the company was floated on the stock exchange amid great media attention. Jaguar had survived the darkest days of the British motor industry, and with the XJ40 project on the horizon, the future looked promising.

XJ-S Specials

Following on from the XJ-S Spyder, Lynx announced the XJ-S Eventer estate in late 1982. Avon had previously carried out a conversion on the Series 3 saloon, but it had been many years since anyone had made a GT into a shooting brake.

It was a fair assumption that some were put off buying an XJ-S because of the lack of carrying capacity, but it was a very brave move adding £6950 plus VAT to a vehicle that already cost £21,000. However, there was nothing else on the market that could offer a combination of near-silent 155mph performance and the same luggage space as a family estate car.

The company's promotional material said: "This conversion greatly enlarges upon the carrying and seating room of the Jaguar XJ-S. Aerodynamically designed, the specially-made side windows give clearer all-round vision; [it boasts] a loading capacity of nearly 42 cubic feet and over 6 foot of load space. More rear seating room is gained by deepening the passenger footwell and moving the rear seats back. This transforms the car into a full four-seater rather than a 2+2. Connolly leather and Wilton carpet are used in trimming the car, exactly matching that of the original."

Having tested the vehicle,

end of 1983, but almost 3000 were on the road by the time that XJ40, the car the engine was designed for, was launched. Given that the six-cylinder XJ-S models were not sold in America (market research revealed that the manual gearbox would not prove popular), these figures were actually quite respectable.

As well as the excitement surrounding the new Cabriolet and 3.6 litre engine at Browns Lane, there were also corporate developments. Admittedly, Jaguar has always been a very introvert company - "like a men's club" as one former employee put it - but it should always have been kept as a separate entity. Coventry Climax, once part of the Jaguar Group, was freed from British Leyland's grip in 1982 (the year Sir Michael Edwardes stepped down from BL) and, at last, the Jaguar-Daimler concern followed two years later, when sales to the USA alone were topping 18,000 units. Jaguar Plc was duly formed and on 10 August 1984,

THE TWO MOST EXCITING ENGINES IN THE WORLD.

THE JAGUAR V-12

In the luxurious street version Jaguar XJ-S an electronic ignition system unleashes the only production V-12 engine in America. The fuel injected aluminum alloy Jaguar V-12 develops 262 HP for a top speed of 139 mph. At normal cruising speed the car has immense reserves of torque for passing or evasive maneuvers. *Road & Track* (May 1984) voted it "Best sports/GT car over $25,000." In its unique blending of exhilarating performance, creature comforts, silence and smoothness, the Jaguar S-type is perhaps the most seductive GT machine in the world.

THE JAGUAR V-12

Jaguars have blazed a trail of glory on the world's legendary tracks from LeMans to Sebring to Daytona. Today, the XJR-5 prototype carries forth this proud heritage. With a modified Jaguar V-12 engine the prototype develops 600 HP for a top speed of over 230 mph. Four victories for the XJR-5 in its first full season of IMSA Racing have added to the Jaguar legend. Thus far in 1984, Jaguars have run at 11 IMSA GT events. The results: XJR-5s have finished in the top three 11 times including a 1-2 sweep of the Miami Grand Prix.

JAGUAR
A BLENDING OF ART AND MACHINE

Another piece of advertising from the States, proving the value of the racing programme from a sales and marketing point of view. As well as enjoying a successful run in the US IMSA GTP series, the Group 44 XJR-5 had just appeared at Le Mans, ending a 20-year absence from the famous 24-hour event, when this advert featured in American magazines.

Spotting a niche in the market, Lynx also decided to build an estate car version of the XJ-S, which it named the Eventer. This advert dates from 1993 - proof that the model was still going strong a decade later - and also shows one of the Lynx D-type replicas.

it was a small price to pay.

Most of the extra horsepower was gained through the use of a high-efficiency stainless steel exhaust system and increased-capacity inlet manifold, which increased output by 10%. Combined with a close-ratio, five-speed ZF gearbox (TWR's ETCC racers used a Getrag unit, incidentally), this gave the TWR XJ-S blistering performance, with 0-60 coming up in just 5.8 seconds, while top speed was said to be 164mph. In keeping with the enhanced performance, the suspension and brakes were suitably uprated, "giving the car impressive stopping power," as one magazine put it.

A less well-known project from the same era was the XJ-S police car. At least one 3.6 litre coupé was prepared for the police at Browns Lane, but, sadly, after extensive trails, it never went into service with the force.

"Jaguar is contemplating joining the designer car business *a la* Lincoln Continentals of recent years," noted *Road & Track*. Burberry was commissioned to trim the XJ-SC in 1984, and two cars duly appeared later in the year featuring the distinctive Burberry check pattern on the hood cover and door pockets; there was also a Burberry-trimmed cloth bag for the targa panels.

In 1985, Guy Salmon, the Thames Ditton Jaguar dealer, produced the Guy Salmon XJ-S Jubilee to celebrate 50 years of trading. Three versions were available, with a miniature replica Jaguar grille, twin-headlights, front and rear spoilers, a rear apron and retrimmed interior. The top model

MotorSport said: "On the road there was nothing untoward about the Eventer's behaviour, no unwanted wind noise and no diminution of the XJ-S's unquestionable performance. Unlike so many estate conversions over the years, the Eventer looks as though it has been conceived as a single, unified design, not as an afterthought adaption."

Continuing the link with the XJ-S (see chapter five) and with the full co-operation of the factory, the TWR concern announced a 330bhp engine conversion, a colour-keyed body kit and special 8.5J x 16 sports alloy wheels for the model. In 1984, it would have cost £34,701 (nearly £13,000 more than the standard V12 coupé of the period), but for the serious enthusiast,

The Lynx Spyder continued to be built in small numbers; this is an HE-based model.

TWR advertising from late 1984.

Turns heads, eats miles

Tom Walkinshaw Racing have used their racing experience to produce an outstanding motor car worthy of any true performance car enthusiast.

However, they have not produced an intractable road-racer, but a docile practical machine that out-performs the original, and stands out in any crowd. High efficiency is the key note throughout, from engine, exhaust system, brakes and suspension, to wheels, tyres and aerodynamics.

This car is as easy on the air as it is on the eye.

Complete cars are available, or we will convert your XJ-S, XJ6 or XJ12 to similar specification.

Data: Power up by 10% to 320 bhp.
Drag down by 12.7%. Lift down by 60% at front, by 88% at rear.
0-60 mph in 5.8 secs. Max speed 164 mph.

For further details, please contact:
TWR Jaguar Sport, Station Field Industrial Estate, Kidlington, Oxford. Tel: Kidlington (08675) 71555.

TWR
JAGUAR Sport
RACE BRED TO IMPROVE THE BREED

THE TWR XJ-S – CLASSIC

THE PARTS.
Altogether, or one by one.

Whether you are considering conversion work on a 3.6 or 5.3 litre Jaguar XJ-S, often the difficult part is knowing where to begin. The TWR Code II conversion will give you the appearance and many of the dynamic advantages you are looking for.

It consists of:

A. TWR Bodystyling Kit. Made to the highest possible standards, the kit consists of front spoiler, rear apron, side sills and boot lid spoiler. Its effect is a reduction in drag and lift, with a tremendous increase in high speed stability.

B. TWR Wheels and Goodyear NCT Tyres, for increased grip and extra cooling for the front brakes. Exclusive to TWR Jaguar Sport, they are also extremely attractive.

C. TWR Suspension Kit. The heart of the machine. Uprated front springs, exclusive gas-filled dampers and revised rear radius arms all combine to transform the ride of the car. More responsive, firmer while still recognisably a comfortable Jaguar. The development came from applying much of the knowledge that was gained while competing in, and finally winning, the European Touring Car Championship.

F. TWR Steering Wheel with thicker, leather covered rim gives you instant control over directional changes with the minimum of effort.

K. Security Wheel Nuts. Sensible to invest a little to guard a lot, 24 hours a day, seven days a week.

L. TWR Power Assisted Steering Valve provides more feel through the steering by reducing the amount of assistance, while still making the car easy and light to manoeuvre, giving feedback to suit the sporting driver.

There is also:

I. If you like your creature comforts, the Interior Retrim might be just what you are looking for. The seat centre panels can be retrimmed in Scottish Tweed, and a no-cost option for the front seats only in the XJ-S is an adjustable lumbar support system using pneumatic bags specially developed by TWR.

J. To set this all off, the chrome Body Trims can be coated satin black or colour coded to match the rest of the car.

D. TWR Brake Kit for those who are looking for extra stopping power. Light alloy four-pot calipers and massive 295 × 35mm discs fit the front, with 270 × 20mm ventilated discs at the rear.

That completes the list of equipment that is suitable for the 3.6 XJ-S. These last five items are only for V12 models.

E. TWR Engine Efficiency Kit is well named. There are larger air intakes, high-flow air filters, special spark plugs and tuned exhaust silencers which are manufactured in heavy duty stainless steel for increased life. This system, together with specific engine settings which are required, produces extra power which is available to the driver from low revs right through the range.

G2. Automatic Gearbox Quick Shift Kit. This package enables the gearbox to respond more quickly to the driver's demands. Whether changing up or down with gear lever or kickdown facility, the modification will allow you to use the power to the best advantage.

G3. Gearshift lever, with round knob in leather or wood, available to fit on Automatic gearboxes, for a finishing touch that recalls the bygone years of classic motoring.

This completes the Code I list, the only remaining items being the optional manual gearbox and 6 litre engine.

G1. 5 Speed Manual Gearbox to replace the 3 speed automatic. Five carefully chosen ratios give you the choice of effortless silent cruising, with a touch over 2000 rpm on the clock at 70mph, or out and out performance driving, where possible, with a maximum over 160mph. Some people believe that a performance car should come with a manual gearbox. We give you that option.

THE WHOLE.
Driving impressions.

We could give you chapter and verse on what it is like to drive a TWR-converted Jaguar XJ-S, but why not let some impartial professionals do that for us?

PERFORMANCE CAR
'Extended side lower panels, a twin pylon rear wing and the treatment of all external brightwork to an anodised matt black finish complete the subtle styling.'

MOTOR
'The revelation is the crisp turn-in to corners where the TWR instantly communicated an assured poise and agility.'
'The handling/ride compromise is cunningly clever. Others made the mistake of stiffening up Jaguars too much. This never feels overtly stiff, just taut and well-controlled enough allow you to quickly forget that you are pressing on in nearly two tons of motor car.'

MOTOR SPORT
'The rate at which the TWR converted model eats up the mile quite astonishing. Like a true performance car it reaches 100mph in around 15 seconds and keeps on going, pulling strongly to 130mph and much more, given the space.'
'The TWR XJS is a car that we'd be happy to match against of its rivals on acceleration, high speed performance, handling and braking.'

FAST LANE
'TWR have turned their Jaguar racing experience to excellent effect with their elegant and indecently rapid six-litre, 5 speed XJ-S.'
'Here is an XJS that looks elegant, whose steering has weight and purpose, that can be driven with . . . precision round corners whose stunning V12 power has been put right at the driver's command through five manual gears. Yet here still remains the silent, gliding ride of the XJS, its effortless performance, its consummate ability to cover ground at high speed.'

ROAD & TRACK
'Walkinshaw's crew doesn't stop at the drivetrain; there is chassis work involved in making this superb sporting car even better. The TWR XJ-S gets beefed-up vented disc brakes front and rear, giving the car impressive stopping power.'
'The ride is quite good at low speeds; the Jaguar by TWR will give you a case of teeth chatter over bumps. But when the speed builds and you start running along at 80, 90, even 100mph or more, the TWR suspension and modifications produce a hunkered-down sort of, well, cat-like grace.'

H. 6 Litre Engine. This is the ultimate in road going Jaguar engines. Each unit is hand assembled side by side with competition engines destined for the race tracks of the world. Using specially forged pistons, a long throw crankshaft (which is machined from solid) to produce added swept volume, and reprofiled valvegear, each engine is dynamometer tested to the specified 380bhp. This superb everyday road engine which will produce vastly expanded power and flexibility and has none of the bad manners of a detuned racing car.

The XJ-S section of the TWR brochure for 1986, describing the many upgraded components and body kits available. Note the TWR treatment was available on the Cabriolet as well as the coupé models.

The Guy Salmon Jubilee model. As the brochure said: "Some of our Big Cats have different faces."

The XJ-S Coupe has always provided a unique blend of effortless performance with Jaguar style and comfort, and the luxury of real burr elm and Connolly leather. With its turbine smooth 5.3 litre V12 engine, the XJ-S H.E. is still unquestionably one of the fastest production automatics in the world.

Now, however, there is another XJ-S Coupe, still with scintillating performance and traditional Jaguar craftsmanship, but having a quite different character from that of the H.E. Coupe.

The raised bonnet channel of the new XJ-S

THE JAGUAR XJ-S COUPE. THE CHOICE IS NO LONGER AUTOMATIC.

3.6 Coupe houses a 3.6 litre, 24 valve, 6 cylinder DOHC engine delivering 225 bhp* through a 5-speed *manual* gearbox, giving a top speed of 145 mph* and an acceleration close to that of the mighty V12: 0-60 mph in 7.5 secs.*

The combination of this sporting engine and slick manual change results in a car ideal for the driver who enjoys responsive performance for its own sake, on motorways, on country roads, at home

or abroad. Even on the Continent you won't be able to stretch the car without breaking the law.

So now you have a choice of Jaguar Coupes. The Jaguar XJ-S 3.6 Coupe. The definitive *sporting* grand-tourer. And the V12 Jaguar XJ-S H.E. Coupe. The ultimate *luxury* grand-tourer.

* Manufacturer's own estimates

JAGUAR The legend grows
JAGUAR CARS LIMITED ENGLAND
THE NEW XJ-S RANGE: H.E., 3.6 COUPE, 3.6 CABRIOLET.

XJ-S 3.6 COUPE £19,495 (MANUAL TRANSMISSION ONLY). XJ-SC 3.6 CABRIOLET £21,495 (MANUAL TRANSMISSION ONLY). XJ-S H.E. £23,995 (AUTOMATIC TRANSMISSION ONLY). PRICES BASED UPON MANUFACTURER'S RRP AND CORRECT AT TIME OF GOING TO PRESS, INCLUDE SEAT BELTS, CAR TAX AND VAT. (DELIVERY, ROAD TAX AND NUMBER PLATES EXTRA.)

(known as the 'Jubilee Special' and costing a massive £40,435) featured the mini grille plated in 24 carat gold, while the bumper trim and door handles were given a gilt finish. The press release stated: "The Jubilee incorporates exterior modifications and interior styling to provide a vehicle of elegance and style which will appeal to the connoisseur of classic motoring taste. It is a car to wear, rather than simply drive."

Sadly, on 8 February 1985, Sir William Lyons, Jaguar's founder, passed away. As Bill Heynes once said: "He was no engineer - he just knew a good car when he saw one." His unique talent would be sorely missed.

The V12 Cabriolet

Midway through the year, on 1 July 1985, Jaguar annouced its first open V12 car since the E-type - the £26,995 XJ-SC HE. With 295bhp on tap in a car weighing 1788 kg, this was every bit as quick as its famous predecessor. Apart from the engine, the new model was much the same as the six-cylinder version; it was still a pure two-seater, and the targa arrangement with a half-length hood was retained.

By now, due to a growing waiting list, Jaguar undertook the fitting of the hood and targa panels. Although Aston Martin Tickford was doing a fine job, its limited production facilities were simply slowing deliveries to an unacceptable level.

Autocar said: "The XJ-S has always stood for effortless performance, and the Cabriolet 5.3 is further proof of this concept. On the one hand, the car

The XJ-SC V12 Cabriolet. From the outside, apart from the badging, it looked almost identical to its six-cylinder stablemate.

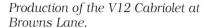
Production of the V12 Cabriolet at Browns Lane.

A V12 coupé from autumn 1985 (UK 'C' registrations started in August that year), pictured at Coombe Abbey on the outskirts of Coventry, England. By now, there were four models in the XJ-S line-up.

is docile enough to be driven round town without showing any signs of discomfort; on the other, it can be used to provide silky smooth acceleration and supercar performance at will. The engine is what sets this car apart from any other on the road - it is superbly refined."

Summing up its road test,

Autosport felt "the XJ-SC V12 is a worthy addition to the Jaguar line-up, and most of the criticisms that can be made of it lie in two areas: the (to us) unnecessary complication of the roof structure, and the fact that, in many ways, it is showing its age. Nevertheless, it is still a highly impressive motor car, and unmatched in the sheer

ease and sensuousness of movement. Others, though, are catching up ..."

A number of minor changes were introduced for the 1986 models, including new badging (the 'HE' badge on the tail was replaced by a 'V12' one), and although elm was retained for the six-cylinder cars, burr walnut was used to trim the V12 interiors. The 3.6 litre models had Herringbone tweed cloth centre sections added to seats, while all cars in the XJ-S range received upgraded stereo systems.

At £26,995, the V12 Cabriolet was just £2000 more than the V12 coupé. The six-cylinder stablemates were priced at £20,395 and £22,395 for the coupé and Cabriolet respectively. There were 17 standard coachwork colours at the time, namely Tudor White, Clarendon Blue, Grosvenor Brown, Cirrus Grey, Black, Sebring Red, Cranberry, Rhodium Silver, Cobalt Blue, Sapphire Blue, Racing Green, Coronet Gold, Regent Grey, Sage Green, Claret, Silversand and Antelope.

In America, where the car made its debut in spring 1986, reactions were mixed, especially as an electric sunroof was now listed as an option on the coupé for just $1300. Although it was the first time the US had been offered the Cabriolet, there was no longer anything new about the format, and from a $41,500 car powered by a 5.3 litre engine, people were - perhaps rightfully - expecting more performance; a 0-60 time of 8.4 seconds was hardly neck-wrenching.

The V12 was also attracting the dreaded gas-guzzler tax by now, which added a hefty $1500 to its price. One

V. Fast.

THE XJR-6. 6.0 LITRE V.12. MAXIMUM SPEED, 230 MPH. AVAILABLE ONLY TO
DEREK WARWICK, EDDIE CHEEVER, JEAN-LOUIS SCHLESSER, AND GIAN-FRANCO BRANCATELLI.

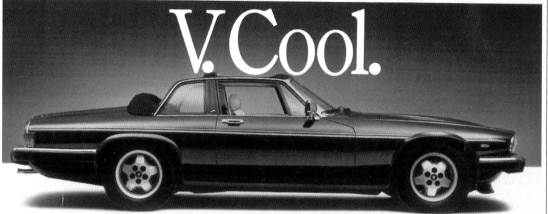

V. Cool.

THE XJ-SC CABRIOLET. 3.6 LITRE 24 VALVE 5 SPEED MANUAL,
OR 5.3 LITRE V.12 AUTOMATIC. AVAILABLE FROM £22,395.

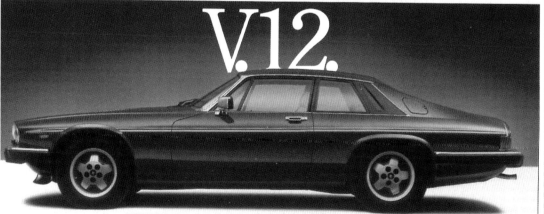

V. 12.

THE DEFINITIVE GRAND TOURER. THE XJ-S 5.3 LITRE V.12 AUTOMATIC,
OR 3.6 LITRE MANUAL. AVAILABLE FROM £20,395.

The XJ-S range of sports and grand touring cars. Four choices. No alternative.

ES BASED UPON MANUFACTURER'S RRP AND CORRECT AT TIME OF GOING TO PRESS, INCLUDE SEAT BELTS, CAR TAX AND VAT. (DELIVERY, ROAD TAX AND NUMBER PLATES EXTRA.)

JAGUAR The legend grows
JAGUAR CARS LIMITED, ENGLAND

British advertising from the 1986 model year. The XJR-6 at the top of the page finished joint third in the 1986 WSPC title chase, only five points down on the winning Porsche team.

Now considered glamorous, the XJ-S was being used in all sorts of advertising. This NGK sparkplug advert shows the XJ-SC HE. Later on, Pirelli even painted a convertible to resemble the markings found on a Jaguar (a real Big Cat, that is!), to promote its tyres.

WHAT MAKES THE CAT PURRRR?

THE JAGUAR XJS V12. JUST ONE OF A RANGE OF TODAYS CARS, FROM LUXURY SALOON TO FORMULA I, PLUS MARINE, HORTICULTURAL AND INDUSTRIAL ENGINES THAT HAVE NGK COPPER CORE PLUGS SPECIFIED AS ORIGINAL EQUIPMENT.

NGK SPARK PLUGS (UK) LTD., 7/8 GARRICK INDUSTRIAL CENTRE, HENDON, LONDON NW9 6AQ. TELEPHONE NUMBER: 01-202 2151/4.

NGK **Copper Core Excellence.**

Although a long while coming, the XJ40 saloon reinforced Jaguar's reputation for fine engineering. It is seen here in Daimler 3.6 guise alongside a rather earlier example of the Daimler marque, to celebrate the Coventry company's centenary.

commentator noted: "It's hard to rationalize a car like the Jaguar XJ-SC. For far less money, there are any number of cars that perform as well as or better than this machine. Few, however, can match the Jag's individuality and panache."

Larry Griffin, writing for *Car & Driver*, said: "As a halfway step toward true open-air motoring, the XJ-SC Cabriolet proves two things: first, major amounts of sunshine can be admitted to the interior of the XJ without sacrificing any of the car's basic good-

ness. And second, although the XJ-SC is a noble and carefully engineered alternative to a steel-roofed coupé, it's no substitute at all for a true convertible."

Equally, *Road & Track* wasn't particularly enthusiastic about the

The 1987 model year 3.6 litre XJ-S coupé for the home market.

Interior of the XJ-S 3.6 for 1987. In the early part of that year, a four-speed automatic transmission became optional on the six-cylinder cars.

Rear three-quarter view of the 3.6 litre coupé, pictured by Chesterton Windmill, just outside Coventry, England. Note the badging on the six-cylinder car.

Cabriolet's "awkward superstructure." Having praised the engine, steering and ride, it summed up the XJ-SC with the following: "If Jaguar returns to a lighter, Son-of-E-type two-seater in the future (an F-type has been rumoured), then *that* might be the more aggressive vehicle that some Jaguar enthusiasts (including us) long for. In the meantime, the XJ-SC is a large, luxurious two-seater, a fine alternative to a Mercedes-Benz or BMW. To keep up appearances, you can wait until you're out of town before opening it up. Then you can put your foot down, *really* open it up and enjoy the sky going by."

However, *Motor Trend* picked up on a very valid point: "Thanks to intensified corporate efforts that brought about huge improvements in the areas

The XJ-SC 3.6 for 1987. The same location was used for a number of the 1987 model year catalogue pictures.

The XJ-S HE pictured with Jaguar's Group C contender for 1987 - the XJR-8. Jaguar went on to dominate the World Sportscar Prototype Championship.

Fascia of the UK specification V12 model for the 1987 model year.

The V12-engined Cabriolet. Note the new coachline for 1987, adopted across the XJ-S range.

of quality and reliabilty, Jaguar has found itself on an especially strong roll in the US for the past several years. We think the arrival of the new XJ-SC Cabriolet will only help continue that saga."

Whilst many car manufacturers were struggling, these were bouyant times for the Coventry company. With production reaching 33,437 units - the best figures in over a decade - profits were reported to be up by 83% in 1985. American sales were still increasing, with 3480 XJ-Ss being sold in 1984, rising to 4861 units two years later, a figure that included 575 Cabriolets.

XJ40

At the end of 1986, Jaguar launched its long-awaited Series 4 XJ6 saloon, better known as the XJ40. From the outset, the car had been designed to take only a straight-six engine, as

Jaguar's engineers didn't want to risk having the V8 Rover unit forced upon them during the darkest days of the BL regime. However, that now meant it would be some time before it could be equipped with the Jaguar V12, so, for the time being, the Series 3 XJ12 and Double-Six saloons continued to be sold alongside the newer models.

Powered by the same AJ6 unit that was used in the smaller-engined versions of the XJ-S, the fact that this new car was available with a four-speed ZF automatic transmission signalled that the 3.6 litre XJ-S should receive it as an option soon. Other interesting features included the outboard mounting of the rear brakes, and availability of ABS.

With XJ40 prices ranging from £16,495 for the XJ6 2.9 to £28,495 for the Daimler 3.6, the XJ-S 3.6 coupé looked good value at £20,995, while its V12 stablemate was also well-priced at

£26,300; Cabriolet models were £2000 more with either engine.

New coachwork colours were announced for 1987, including Black, Grenadier Red, Jaguar Racing Green, Nimbus White, Westminster Blue, Alpine Green, Arctic Blue, Bordeaux Red, Crimson, Dorchester Grey, Moorland Green, Satin Beige, Silver Birch, Solent Blue, Sovereign Gold, Talisman Silver and Tungsten.

Having mentioned British Leyland, just for the record, it should be noted that BL was renamed the Rover Group in 1986; two years later, it was bought by British Aerospace, which, in turn, sold it to BMW in 1994. Jaguar's fortunes were thankfully somewhat better. The XJ40 gained numerous awards for Jim Randle (the car's Chief Engineer) and his team and sales across the Jaguar-Daimler range continued to rise.

Despite admirable sales figures,

108

The Daimler-S pictured outside the old Limousine Shop, where it was built. Note the two-piece targa roof painted in body colour, and the different lines made possible by deletion of the flying buttresses. The car went to a styling clinic in the States, but was rejected.

Detail shot of the handmade Daimler-S grille, complete with traditional fluting.

The fluted bootlid and badging at the rear of the Daimler-S. Thankfully, the car was retained by Jaguar, and now forms part of the JDHT collection.

helped, no doubt, by a number of accolades and a string of favourable road test reports in the USA, the company was still looking into ways of broadening the appeal of the XJ-S. In 1986, Keith Cambage's limousine shop built the Daimler-S prototype. It was displayed at a clinic in America, but sadly, never went into production as it was felt demand (particularly in the States where the Daimler name is still relatively unknown) would not be high enough to justify the added manufacturing costs. In addition, Jim Randle had a one-off, long-wheelbase coupé - interestingly, also minus the flying but-

The 1988 model year S in suitable surroundings (just outside of Le Mans), given Jaguar's success in the World Sportscar Prototype Championship.

tresses - produced, but this, too, was destined to be a stillborn project.

More changes

Four months after the launch of the XJ40, it was announced that the 3.6 litre XJ-S could at last be bought with an automatic gearbox. Priced at £740, this was the same four-speed unit as employed in the saloons, and would enable the coupé model to break through the standing-quarter in 16 seconds without any effort on the part of the driver; 0-60 could be covered in 7.8 seconds. Strangely, the highly-praised 'J-gate' found on the saloons was not adopted on the Grand Tourer, despite its more suitable application for added driving enjoyment. All cars in the XJ-S range gained redesigned switchgear, heated door mirrors and stainless steel treadplates.

In summer 1987, Jaguar disclosed it had just produced its 100,000th V12 engine and 50,000th XJ-S. Given that the futures of both were extremely uncertain at the beginning of the 1980s, this is an admirable achievement which stands as testiment to the engineering

The 1987 Championship-winning XJR-8. The Silk Cut Jaguar team ended the season with almost double the points of its nearest competitor.

The 1988.5 model year V12 coupé, as seen in the UK catalogue.

XJ-S V12 COUPE

Powered by the same race-proved 5.3 litre engine as the Convertible, the XJ-S V12 Coupe combines smooth power with the highest standards of personal comfort and luxury.

With equipment and fittings similar to those of the XJ-S Convertible, the V12 Coupe also shares the sophisticated anti-lock braking system in which each wheel is individually monitored by a sensor which sends reports on the status of the wheel to a central electronic control unit six times every second. The microprocessor instantly analyses this and other information, then signals increases or decreases of hydraulic pressure to the relevant disc brake.

The self-checking anti lock braking system also features 'yaw' control, which automatically compensates for varying rates of braking, as – for instance – when the near side wheels are on a wet surface and the offside ones are on a dry one.

The sporting yet elegant appearance of the Jaguar XJ-S V12 Coupe accurately reflects its character, in which race-bred power and precise handling are teamed with unique standards of comfort and refinement.

Wheel trims may differ from those shown [illegible]

The 3.6 litre XJ-S in the same publication.

XJ-S 3.6 COUPE

The sporting nature of the XJ-S 3.6 Coupe stems from the integrated design of its power unit, transmission and chassis components, the firmly locating seats and twin-spoked steering wheel. In particular, the sports suspension system developed especially for the 3.6 Coupe is perfectly complemented by the wide-rimmed 6.5in spoked alloy wheels fitted with 235 x 60 low-profile tyres and servo-assisted anti-lock braking.

The 3.6 litre 24 valve 6 cylinder double-overhead-camshaft engine with its electronic engine management system, develops 221bhp (165kW)*. This is transmitted through a 5-speed Getrag manual gearbox, for those preferring automatic transmission, a 4-speed ZF automatic gearbox with power-preserving 4th gear lock-up is available as an optional extra.

Correctly designated a 2 + 2, the 3.6 Coupe has comfortable occasional rear seats, trimmed like the front seats in hand-stitched cloth and leather. Traditional burr elm veneer and generous standard equipment such as central locking, electric windows and full air conditioning, all contribute to the luxurious nature of this sporting car.

* Manufacturer's estimate

The XJ-S V12 Coupe has the potency to bring a distant blur into sharp focus.

Effortlessly.

Equally impressive is an advanced anti-lock braking system like no other in the world.

Developed over many years by a team of engineering perfectionists, it even has anti-yaw control.

Together with an anti-dive independent suspension system, it means the V12 Coupe can take varied road conditions in its stride.

While to keep the outside world in its place air conditioning is standard equipment.

Although the heated front seats are a totally new bolstered design, some things remain unchanged.

If you still seek that unique combination of wood and hide, you will not be disappointed.

Leather even stretches to the steering wheel.

In the XJ-S V12 there are certain traditions we would never break.

JAGUAR · THE V12 COUPE · XJS · WORLD SPORTSCAR CHAMPIONS

JAGUAR CARS LTD, COVENTRY ENGLAND

This particularly tasteful piece of British advertising for the V12 coupé dates from spring 1988.

abilities of the Coventry concern.

At the Frankfurt Show in September 1987, a number of improvements were announced for the XJ-S range. The 3.6 coupé featured a bolder twin coachline, new alloys, sports seats and a thicker-rimmed steering wheel, while the V12 models, as well as inheriting the new seats and steering wheel, gained leather trim on the centre console, armrests and door pulls.

From a mechanical point of view, the 3.6 litre AJ6 engine received some attention with the aim of improving refinement. *Autocar* noted: "The all-alloy 3.6 litre engine has undergone a lot of careful development over the years, and the twin-overhead camshaft, four-valves-per-cylinder unit is now as smooth and refined as the best - certainly there is no trace of the on/off idle roughness of early examples and the AJ6 unit only begins to sound busy as its 5700rpm redline is approached. The microprocessor-controlled ignition/fuel-injection system is a Lucas/Bosch unit integrating the two functions according to the obligatory 'maps' in its memory. It seems to have been responsible for a slight loss in peak power, down from 225bhp at 5300rpm to 221bhp at 5250rpm, but peak torque is up 8lbft to 248lbft."

With the objective of providing the smaller-engined model with a truly sporting character, the 3.6 coupé also received some significant modifications in the chassis department; dubbed the Sports Handling Pack, there were different anti-roll bars, spring rates were raised, more 'feel' was put into the steering, and new Pirelli 235/60 VR15 low profile tyres were specified. Priced at £23,821 on the road, the 3.6 litre XJ-S could now compete head-on with cars like the Porsche 944.

The Jaguar was not only fractionally cheaper, it was quicker, too. At the end of a comparison test, *What Car?* concluded: "At the finishing post, it's a win on points for the Jaguar, not a knock-out victory as some predicted at the outset. In the final analysis, the Jaguar's healthy price advantage, superior performance and space are factors that are hard to ignore. Its attractive wood and leather interior treat-

The XJ-S Convertible pictured at the time of its launch. Direct glazing on the windscreen and removal of quarter-lights from the front side windows helped improve aerodynamics to a point where the Cd was almost the same as for the closed coupé.

ment appeals too. It's a big car, the XJ-S, less nimble perhaps than the Porsche and not as sexy looking but still, for all that, a highly desirable proposition. The 944 is good, but for the money, the Jaguar is better."

After its launch, the V12 Cabriolet proved far more popular than the six-cylinder version, so the latter was discontinued for 1988. By October 1987, the XJ-SC was being listed in V12 form only at a price of £31,000. Princess Diana owned one for four years, and was often seen on the television driving the car. Her particular example carried a number of detail changes, including back seats for Prince William and Prince Harry.

The home market used to take around half of XJ-S production, but now exports were accounting for around three-quarters of XJ-S sales - a record 5380 found new homes in the US in 1987, with only 597 less in 1988. This not only reflected America's new-found faith in Jaguar's reliability and build quality, but also the fact that the various models in the S line-up were more suitable for this, the world's largest outlet for sporting cars.

This was a very good thing, as Black Monday (the massive stock market crash of October 1987) would seem to indicate a rocky future for luxury car makers on the home market. In the background, the American car giant, Ford, bought Aston Martin, the Newport Pagnell-based prestige sports car builder that, for many years, symbolized the epitome of British craftsmanship with regard to the automobile. At the time, few could have seen the rel-

evance of this purchase, but it had some interesting consequences.

The 1988.5 model year changes came in February 1988, when the Teves anti-lock braking system was fitted as standard across the XJ-S range. The V12 also gained heated seats with electrically-controlled lumber support, and front fog lamps. However, a new S was about to be launched in Geneva, exactly 27 years after the E-type had stunned the crowds gathered there.

The Convertible

Visitors to the 1988 Geneva Show were given a preview of the stunning XJ-S Convertible. Powered by the 5.3 litre V12 engine, this beautiful model replaced the Cabriolet in the Jaguar line-up. Priced at £36,000, it was the most expensive Big Cat to date (only the Daimler DS420 limousine carried a higher price tag, listed at £40,000 in standard form), but it was still exceptional value for money, and indeed Sir John Egan later admitted "we may have under-priced it."

Ever since the birth of the S, there had been plans for a drophead version, known at that time as XJ28. However, it wasn't until mid-1985 that the idea was officially revived. Free from Leyland's grasp, and therefore able to

tackle the soft-top at last, on Egan's instructions, a Project Team consisting of 12 members was assembled. Headed by Ken Giles, it included staff from both the engineering and sales and marketing sides of the company: it was imperative that Jaguar got this car absolutely right as it called for a great deal of investment.

Giles recalled: "The Chairman's brief to us was typically straightforward - he wanted a world class convertible, we had to have a saleable vehicle on our stand at the 1988 Geneva Show, and be able to start selling cars in the UK and Europe by May 1988."

Although the convertible employed the coupé floorpan, there was much more to the new model than meets the eye. In fact, in order to retain structural rigidity and eliminate any chance of scuttle shake, this was a completely re-engineered vehicle.

Under Jim Randle, and with the help of the Structural Dynamics Research Corporation, early design work was subjected to Dynamic Finite Element Modelling (DFEM). This involved the use of computer programmes to highlight problem areas instead of having to go to the expense of building prototypes. Using this information, a

With the hood up or down, the XJ-S V12 Convertible is a classic Jaguar embodying the highest standards of progressive engineering and traditional craftsmanship.

Technology is exemplified by the smooth twelve cylinder engine, with its electronic fuel injection system, and the advanced anti-lock braking system. This technical mastery is uniquely complemented by a wealth of hand-crafted refinements, supple leather and polished walnut.

Effortless high-speed cruising is complemented by responsive handling and confident roadholding; beneath the Convertible's style comfort and luxury is half a century of sporting achievement: a heritage exemplified in the race-bred V12 OHC engine capable of developing 291bhp (217kW)*.

Just about every conceivable aid to motoring pleasure is here, including central locking, cruise control, trip computer, inlaid walnut veneers and hand-crafted leather upholstery, air conditioning and heated seats with electrically adjusted lumbar support.

The concept of a completely open-topped Jaguar, last seen in the legendary E-type, is gloriously reborn in the new XJ-S V12 Convertible. Here all the power, precise handling and elegance of a Grand Touring car are enhanced by a tailored hood of unique design that opens and closes at the touch of a button.

As the hood folds back, pillarless rear windows glide out of sight, affording the driver and passenger the air of freedom which can only be experienced in a truly open sports car.

Operating the hood takes only twelve seconds. It stows neatly behind you (a padded cover is provided) and the rear quarter windows open simultaneously. With the electrically-operated front windows also down, you have a completely open car.

Fitted with a tinted and electrically heated glass rear window, the hood is fully lined and padded to provide a high level of insulation and is hand-tailored to obviate draughts.

Structural rigidity is a vital consideration in open car design and the XJ-S V12 Convertible chassis has been selectively strengthened in the light of extensive finite element studies. The result is exceptional torsional stiffness which even exceeds by a considerable margin the stringent targets set by Jaguar engineers.

The new Convertible as it appeared in the first UK catalogue to feature the model; it was given two full double-page spreads. It is interesting to compare the quality of this production with the rather quaint original of 1975.

mock-up was built and analyzed on a test rig before Karmann was asked to produce fully-engineered prototypes for an extensive testing programme.

No less than 108 new panels and 48 modified pressings were employed in the construction of the Convertible, with reinforcement in the transmission tunnel, front and rear bulkheads and rear floor area; in addition, steel tubing ran through the sills and windscreen pillars. Karmann produced the tooling for the new panels, whilst the pressings came from the Rover Group's factory in Swindon, but this turned out to be a worthwhile investment - the soft-top model had only 15% less torsional rigidity than the coupé.

Karmann was also responsible for the design and engineering work of the hood mechanism. Two catches needed to be released on the screenrail and then a switch took care of the rest, electric motors folding back the hood until it sat neatly in the recess behind the seats - like the now-defunct Cabriolet, the Convertible was a strict two-seater. The Porsche 911 Carrera Cabriolet was the only other European sports car to have a power top at that time, and in keeping with Jaguar's quest for reliability, the hood underwent a very severe test programme.

Despite extra strengthening in the body and all the mechanism for the hood and rear quarter windows (which operated in tandem with the hood), at 1900kg in UK trim, the Convertible was only 100kg heavier than the V12 coupé. The 0.39 Cd figure was only 0.01 higher than that of the closed car, so, as one would expect, performance was very similar.

The European launch was on the Cote d'Azur in France. Shortly afterwards the cover of *Motor* featured the drophead XJ-S with the headline: "Jaguar Convertible - driving the best XJ-S yet." Headings like that became commonplace, as virtually every journalist was bewitched by the new model. "A real cad's car but one with long-legged performance, a car you could drive for the sake of it," said Eoin Young. UK sales began on 27 April 1988.

In the US, there already was a convertible, of course. The Americans had requested a drophead on a number of occasions, but when told how long they would have to wait, the people in New Jersey decided to commission Cincinnati-based Hess & Eisenhardt (the company appointed by Jaguar Cars Inc. to fit the S-type's official sunroof option) to build one for them.

Announced just in time for the 1987 model year (at the Miami Show on 29 October 1986), the H&E conversion included an electrically-operated hood and a substantial number panels to strengthen the bodyshell. Built to order at a cost of $47,000 apiece, it was, sadly, only ever sold in the States. Naturally, when the factory model became available, the Hess & Eisenhardt arrangement came to an end; between 1986 and 1988, around 2000 examples were produced.

In addition to the Hess & Eisenhardt convertible, which was backed by Jaguar Cars Inc. in North America, another US-built soft-top made its debut in time for 1987. The Eurowest concern of California offered new cars for $57,000, or could carry out a conversion on a customer's vehicle for $16,000, with the work typically taking a month to complete.

But now, at last, a car was available from Browns Lane. Regarding the Coventry-built soft-top, *Motor Trend* said: "You'd naturally expect a high level of finish in a car that costs about 30 grand per seat [it was launched at $56,000 at a time when the coupé was listed at $47,000]. The Jaguar XJ-S convertible won't disappoint you ... True to its character, the Jaguar feels composed and relaxed under all circumstances. The assisted steering has an accurate, substantial feel. Fully independent suspension strikes a fine balance between handling crispness and ride quality; the car never misbehaves, even when confronted with major blemishes in the pavement, and little of the impact makes its way past the suspension and carefully tuned

A glamorous setting for a glamorous car - British advertising for the new Convertible.

The factory-built, American specification Convertible, distinguished by the front side marker and auxilliary high-level rear brake light mounted on the boot. It's noticeable, too, that the bumpers are slightly thicker than on European models, which added 100mm (4in) to the overall length.

bushings. Even at a frisky pace through turns, the Jaguar feels balanced and completely in control. The Pirelli P600s stick nicely but also transmit little road noise into the chassis - a necessity in a car of this class. And should the road deliver an unpleasant surprise around the next bend, the Jaguar offers the added security of anti-lock braking."

Road & Track was equally impressed with the refinement: "Top up, our test car was quiet and draft-free, except for a small whistling wind leak at the left rear window. Otherwise there's no flutter, flap or other evidence that you aren't in a nice solid fixed-head coupé, even at high speed. With the top down, wind flow over the windshield is remarkably smooth with

a near absence of turbulence in the cockpit. Whether at 25 or 75mph, the wind merely tousles the hair, like a light studio fan in a back-shot Hollywood movie. No horizontal neckties or sunglasses lost overboard.

"It just may be the most handsome XJ-S yet produced. The shorter roofline accentuates the length and gracefulness of the hood and trunk,

117

The V12 engine that powered the Convertible. Subtle changes can be spotted if one compares this unit to the earlier HE engine pictures. (Courtesy Paul Skilleter)

A final shot of the Convertible at the time of its launch.

giving the car that low, elegant cat-like look for which Jaguars have always been famous."

The Convertible was an immediate hit Stateside, accounting for 57% of the 4458 XJ-S sales in 1989, and sales continued to rise the following year; 1990 was the best year for US V12 Convertible sales, with 3057 examples sold.

In the meantime, from December 1988 the V12 engine was fitted with Marelli digital ignition (there were no changes in either power or torque), and a Sports Suspension Pack was listed as an option for the V12 coupé. By that time, the 3.6 litre coupé was priced at £26,400 on the home market, whilst the V12 version was listed at £32,000, the soft-top model being £6000 extra.

JaguarSport

On 9 May 1988, the JaguarSport organisation was founded as a 50/50 venture between Jaguar and TWR. Tom Walkinshaw was declared Managing Director, with Sir John Egan Chairman. Egan said: "The formation of JaguarSport is a logical development of the highly successful motorsport relationship Jaguar has established with TWR over the past few years, which has brought us the European Touring Car and World Sportscar Championships, and victory at Le Mans. It will enable us to apply the experience we have gained on the racing circuits of the world, to develop Jaguars aimed at the more specialised requirements of the enthusiast."

The first model appeared in late August - the £38,500 XJR-S - with the first batch of 100 finished in Tungsten (a metallic mid-grey) with doeskin leather trim, and numbered to celebrate Jaguar's WSPC success and its victory at Le Mans that year; the

THE XJRS V12
CELEBRATION
COUPE

SPORT

The XJR-S. The first 100 cars were given the 'Celebration Build' appellation, and were finished in a unique colour scheme.

treadplates featured the build number, winning laurels and the JaguarSport logo. After this initial run, there was a choice of four colours - Signal Red, Solent Blue, Alpine Green and Arctic Blue, all with magnolia trim.

The bodywork, designed by Peter Stevens (who came up with the styling on the new Elan for Lotus, launched at the end of 1989), reduced drag and lift by around 10%, and although the engine was left in standard trim, suspension and steering were uprated to give the car a more sporting character. The finishing touch was a set of 7.5J x 15 inch Speedline wheels shod with 235/60 VR-rated tyres.

Despite the fact that the changes were quite modest - stiffer springs, Bilstein shock absorbers, a thicker anti-roll bar, harder bushes, a change in the steering ratio and a few slight adjustments to the geometry - the difference was very noticeable. As *Fast*

The Jaguar stand at the 1989 Geneva Show, the event at which Mercedes unveiled its all-new SL.

Jaguar's other supercar, the magnificent XJ220. After going into limited production at the JaguarSport facility, a few found their way onto the race tracks. David Brabham, John Neilsen and David Coulthard took the car to a GT class victory at Le Mans in 1993, but it was later disqualified for a highly questionable technical infringement.

Lane pointed out: "Even with a standard engine, the XJR-S felt a lot quicker simply because it gave more traction and was less turbulent on uneven roads."

Production was set at 500 cars a year, rising to 2500 according to customer demand. As Sir John said at the Waldorf Hotel launch: "There aren't really any cars like the XJR-S already in the market. The important thing is that we know the demand for it is there. It's like the XJ-S Convertible - we knew there was demand for it, and we proved we were right. We sold 1000 of them in the first few days." But the financial climate was about to take a turn for the worse. Not only would it affect cars like the XJR-S, but also another Jaguar debutante - the XJ220.

The stunning XJ220 was presented to the public at the NEC in October 1988. As *Autocar & Motor* said at the time: "There, naked in the lights, stood the final proof that Jaguar had the confidence and know-how to build a car that could be mentioned in

Jaguar's promotional material just got better and better. This American catalogue, dated July 1988 and introducing the 1989 model year range, was beautifully produced. Here we see America's 1989 MY Convertible, complete with matching XK120.

Front view of the US specification coupé for 1989. At this stage, twin headlights were still used on all XJ-Ss shipped to the States, and passive seat belts had been introduced to meet that market's requirements.

Below: A rear view of the same car. Note the high-level rear brake light.

petition from Stuttgart and fears of a recession. No less than 11,206 were produced - an all-time record for the XJ-S, and one, incidentally, that would not be beaten.

Other Specials

Brian Lister has been linked with the Coventry marque since the 1950s, when Jaguar-powered racing cars bearing his name were campaigned by the likes of Archie Scott-Brown, Stirling Moss, Jim Clark and Ivor Bueb. It was

the same 200mph breath as the Ferrari F40 and [Porsche] 959."

This incredible piece of craftsmanship caught the imagination of everyone - a mid-engined, four-wheel drive supercar with its highly modified V12 engine developing 500bhp. The 220 designation was chosen to symbolize the car's top speed - a staggering 220mph. Eventually, JaguarSport put the car into limited production with a turbocharged 3.5 litre V6, but as the recession bit, converting deposits into sales became a difficult job.

However, generally speaking, and no doubt helped by an excellent performance in the World Sportscar Championship, Jaguar was riding high; XJ-S sales at last passed the 10,000 cars per annum mark in 1988.

Mercedes-Benz launched the all-new SL at the 1989 Geneva Show, but projections for 1989 were extremely accurate. Jaguar expected to build 12,000 XJ-S models, of which 5000 would be convertibles - at the end of the year, they were well within 7% of these estimates, despite the new com-

therefore only right that Lister should turn its attention to the XJ-S.

Lister enlisted the help of Ron Beaty (an ex-Jaguar development engineer), John Lewis and Iain Exeter, and BLE Automotive was formed to market Lister products. From 1984, various body styling kits and interior parts were available, but it was Lister's work on the V12 engine that created the greatest interest. A seven-litre version was available (later, it could even be bought with twin-superchargers to give 604bhp), but it was the six-litre conversion that proved the most popular.

By lengthening the stroke to give a 5995cc capacity, and carrying out a number of modifications on the induction and exhaust system, fitting new pistons, cylinder heads and camshafts, power was increased to an outrageous 482bhp at 6000rpm! The torque figure was also vastly higher than for the standard engine, with 490lbft at 4250rpm being quoted. The 0-60 time was cut to just 5.4 seconds, while top speed was now around 170mph.

Naturally, with performance like this, the suspension (each suspension conversion was carried out to the customer's individual specification), braking system and steering were suitably uprated. Many opted for the five-speed manual gearbox and modified lsd options, and, of course, the package could be finished off with a range of wheels and tyres.

The Lister XJ-S was also sold in America, known as the Lister XJ-S NAS (the NAS standing for North American Specification). Due to tight Federal regulations, the engine received only mild modification, but fitting the UK specification Getrag gearbox and corresponding 3.54:1 final drive made a significant difference to performance, nonetheless.

Even in 1987, a Lister XJ-S wasn't cheap in the UK - craftsmanship never is - but by 1995, even with the customer supplying the donor vehicle, a full Lister Mk.III conversion cost a hefty £53,505 plus VAT! By that time, Listers were sold through WP Automotive, and the aggressively-styled Le Mans Convertible had also joined the line-up.

A number of well-known tuners worked their magic on the six-cylinder XJ-S as well. Janspeed, one of the country's leading companies in the field of aftermarket turbocharging, produced a twin-turbo package for the AJ6 engine. The XJ40 saloon was the first to receive this installation, but then a similar conversion was offered for the 3.6 litre XJ-S coupé. Power increased to 295bhp (virtually the same

The Lynx Performer brought V12 performance - and often more - to the 3.6 litre coupé.

THE JAGUAR XJ-S JUNIOR. A SMALL MIRACLE.

£3,950 plus VAT, includes number plates.

JAGUAR

Further details on the new Jaguar XJ-S Junior can be obtained from your local dealer or by writing to: Complete Communications, Unipart House, Cowley, Oxford OX4 2PG

A faithful half-scale reproduction of the XJ-S V12 Convertible, the Jaguar XJ-S Junior is proof positive that the best things also come in little packages. Drivers in the 4 years to pre-teens range will find that this latest addition to the Jaguar line echoes the power and grace of the original. Proven technology is used in the tried and tested steering system and in the electronic braking / acceleration system, which is supplied by a 24 volt motor and controlled by a single pedal to reach a maximum speed of approximately 12 mph.

The interior is typically luxurious, upholstered in genuine Connolly hide and carpeted throughout. From the wood veneer dashboard, the driver can operate the working headlights, tail-lights and indicators as well as the stereo radio cassette built in for added pleasure.

Available in Signal Red, Glacier White or Racing Green, the XJ-S Junior wears the authentic Jaguar badging with the distinction of one born to win.

A CHILD PRODIGY

Perhaps the most novel XJ-S special was this, the XJ-S Junior. Built at half-scale and powered by a 24v electric motor, the Junior had the same polished wood and Connolly leather as the original, along with a stereo and a full set of working lights. Governed to 12mph, it could be bought in left or right-hand drive, and came in green, red or white.

Interior of the last of the 3.6 litre coupés. Note the thick-rimmed steering wheel.

ing kit that actually works: the rear aerofoil alone giving extra downforce for every 1mph travelled. For driver comfort, control and safety, Lynx offers its own full electrically-operated competition style seats."

Eventually, Lynx offered three stages of tune: the 280T (which developed 280bhp after a turbocharger was added to a fairly standard AJ6 engine). The other two options, the 350T and 450T, were a bit more involved, calling for a free-flow exhaust system, new pistons, camshafts and injectors, and on the 450T, an increase in the bore to give a four-litre capacity. The latter produced 450bhp and something like 500lbft of torque - the same as a Chrysler Viper GTS!

Towards the end of 1989, William Towns (the designer behind the gorgeous lines of the Aston Martin DBS and the wedge-shaped Lagonda saloon, introduced at the 1976 Earls Court Show) displayed his Railton convertible - a rebodied XJ-S. With aluminium body panels and typically-original styling, it was priced at a staggering £88,775. Nonetheless, Towns was still confident of selling one car a month, and said it was "the sort of car that Sir William Lyons would be building were he still alive today." Even this figure proved somewhat ambitious ultimately.

The Ford takeover

The four-litre XJ6 range made its debut at the 1989 Frankfurt Show. The 3980cc version of the AJ6 developed 6% more power than the original 3.6 litre unit but, more significantly, 13% more torque. To complement this new

as the standard V12), but the maximum torque figure shot up to a stump-pulling 340lbft at just 4000rpm.

Having transformed the XJ-S into both a Spyder and a shooting brake, Lynx was quick to offer a performance package for the 3.6 litre coupé. The publicity material stated: "The Lynx Performer delivers racing performance to the long-legged XJ-S by providing many states of engine tune, including

special six-branch inlet and exhaust manifolds, boring up to 4.3 litres, turbocharging and many other options.

"To cope with the enhanced performance, Lynx has also developed an uprated suspension and braking system, including a very effective rear wishbone conversion that adds greatly to stability in fast cornering.

"To compliment these mechanical modifications, Lynx offers a body styl-

The Convertible and V12 coupé for 1990. By now, the 5.3 litre XJ-S had established itself as the definitive Grand Tourer. Even at £34,200, it represented excellent value when one considers the £57,267 asked for a 928 S4 Porsche, or the £63,200 price tag attached to the Mercedes-Benz 560SEC. The 412 Ferrari (an updated version of the 400i) cost over £80,000!

and much-improved engine, a new ZF four-speed, electronically-controlled, automatic gearbox was employed, featuring a switch that enabled the driver to choose between 'Normal' and 'Sport' settings.

At about this time, news of a potential takeover bid was brought to the public's attention. The poor exchange rate meant that Jaguar's crucial sales performance in the States was seriously weakened, and several large concerns let it be known that they were interested in the Coventry company, General Motors, Ford, BMW and Mercedes-Benz in particular.

Eventually, however, on 1 December 1989, Jaguar shareholders ap-

Before you experience our new 6 litre, appreciate what's behind it.

Behind the new JaguarSport XJR-S 6.0 litre is a history of high performance: the now-classic C Type, D Type and the racing version of the XJ-S; their collective track record making the marque the Legend it is today.

To reflect its racing pedigree, the first 25 of the new XJR-S 6.0 litre are finished in classic Jaguar Racing Green.

Reassuringly, the interior still has that timeless blend of hand-stitched hide and rich polished veneers. But elsewhere, all is progress.

The JaguarSport 6.0 litre V12 engine delivers 318bhp at 5250rpm and a maximum torque of 362 lb/ft at 3750rpm. 0–60 is achieved in a fleeting 6.5 seconds. Mid-range performance is outstanding, with a top speed approaching 160mph (where conditions allow).

Speaking of performance figures, the collective value of the racing legends above currently stands at £1.75 million. Much appreciated Jaguars in every sense.

The new XJR-S 6.0 litre will undoubtedly take its place alongside the great Jaguars of the past.

Arrange a test drive at your nearest JaguarSport dealer now. And experience history in the making.

For details of your nearest dealer please contact JaguarSport Limited, 17 Station Field Estate, Kidlington, Oxford OX5 1JD. Tel: (0865) 841040. Fax: (0865) 841170.

SPORT
A Passion for Performance.

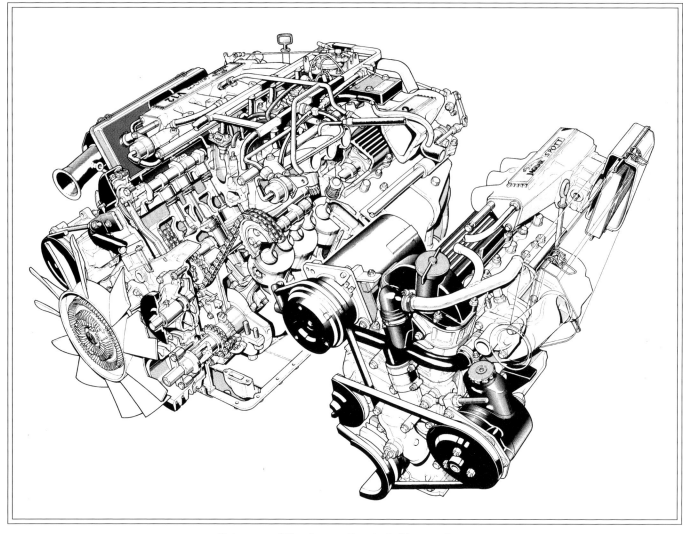

Cutaway of the JaguarSport six-litre engine.

proved a Ford takeover and the government duly waived the so-called 'Golden Share'. Ironically, Ford had approached Jaguar some years earlier with a view to producing a badge-engineered luxury car for West Germany and the USA. With Daimler-Benz (Mercedes) in Germany, there was little point marketing a British Daimler in that country (although a few were sold) and Daimler never really made an impression on the US market. Ford therefore suggested that the Lanchester name be revived, but market research showed the marque had lain dormant for too long, and negotiations broke off at an early stage. Now, though, Ford had

complete control of the Coventry car maker.

The XJ-S into the 1990s

On 23 August 1989, JaguarSport announced the six-litre version of the XJR-S coupé. Priced at a whopping £45,500, this model was a true high performance machine. As *Automobile* magazine pointed out: "Jaguar's XJ-S is becoming long in the tooth - the E-type's sort-of-replacement was launched way back in late summer 1975 - but its claws have been sharpened and its sinews stiffened. In other words, the wonderfully refined V12 engine has been modified to

produce significantly more power and torque. The new six-litre's extra muscle is complemented by other key features, notably an uprated suspension package engineered to make the long, luxurious, and very heavy 2+2 coupé more of a crisp, responsive sports car and less of a fast-but-soft cruiser that tends to be driven with just a finger

Right: Japanese advertising from summer 1990. Sales have increased consistently in this difficult market where Mercedes-Benz and BMW have tended to dominate the luxury import scene for a number of years.

128

アントレプレナーの、J。

さて、ジャガーXJ-Sという車は、一筋の勇気、または度胸を乗り手に要求する。周囲の視線を浴びて、毅然とした態度と優しい眼差しを示せる、男たち。そういう、少しオーバーに言えば、人生の修羅場をくぐってきた、そしてビジネスの最前線に立つ起業家精神の持主、いわばアントレプレナーに、かくも似合うのである。精悍。重厚。繊細、かつ大胆。そんなプロフィールがこの車の、どこか非凡なプロポーションとフォルムに重なって、その魅力は、余人をもって代え難い。ときに、中世のイギリスを発祥とするゴルフというスポーツは、周知のとおりプレーヤー自身が審判員である。正義感とフェアプレー精神に立つ、たゆまぬ自己への研鑽は、この車のオーナーならではの名誉であり、また冥利であろう。だからこそ今、XJ-Sを駆る、しなやかな男たちは、Mr.Jと称賛されるのだ。アントレプレナーの、J。

信頼の品質とサービスを、お届けします。ジャガージャパン

JAGUAR
知るほどに、ジャガー。

Jaguar Japan Limited

and thumb on the steering wheel."

The 5993cc capacity was achieved by increasing the stroke, and with the Zytek sequential injection and digital ignition management system, the engine now developed a healthy 318bhp at 5250rpm and a mighty 362lbft of torque at 3750rpm. As a result, official JaguarSport performance figures claimed a 0-60 time of 6.5 seconds and a rev-limited top speed of 158mph - more than adequate for even the most enthusiastic driver.

Even an experienced tester like Phil Llewellin was moved to say: "Jaguar's challenger is not categorized as a sports car in my book. But the JaguarSport treatment has made one of the world's most delectable *gran turismos* even more adept at reeling in the miles. A few more horses, and a lot more torque, make the XJR-S a terrific mid-range performer. The stretched V12 generates just a little more noise than its virtually silent XJ-S counterpart, but whispering, cream-and-silk smoothness belies acceleration strong enough to make a Trappist monk break his vow of silence to express astonishment and delight." In *Jaguar Quarterly* (nowadays known as *Jaguar World*), Paul Skilleter, the world's foremost authority on all things Jaguar, stated: "Quite simply, it is the best XJ-S yet."

As for the standard cars in the XJ-S range, in October 1989, the 3.6 litre coupé was priced at £27,200, whilst the V12 equivalent was £34,200 (a hefty £2200 more expensive than it had been at the same time the previous year). By this time the Convertible was £41,200 - although this seemed a high price, it actually represented excellent value when compared to the £49,300 price tag attached to the Ferrari 328GTS, but it was only £500 cheaper than a basic Mercedes 500SL. However, to get the SL to a similar specification as the XJ-S would be a very pricey business!

Coachwork colours for the time included Signal Red, Black, Grenadier Red, Jaguar Racing Green, Glacier White, Westminster Blue, Diamond Blue, Bordeaux Red, Regency Red, Savoy Grey, Jade Green, Satin Beige, Arctic Blue, Solent Blue, Gunmetal, Talisman Silver and Tungsten.

In America, the Collection Rouge XJ-S was introduced on 16 June 1989 to augment the line-up. This featured Signal Red bodywork with a gold coachline, lattice alloy wheels with red highlights (now shod with 235/60VR15 tyres, standard across the XJ-S range for 1989), and a magnolia leather interior with red piping - the magnolia even extended to the steering wheel and gearknob. Priced at $51,000, it was $3000 more than the standard coupé, and $5000 less than the Convertible.

Writing for *Road & Track*, Richard Homan summed up the way in which the XJ-S had matured and grown to be accepted with the following prose: "The XJ-S is a self-assured automobile with breeding, class and panache. And without pretentions. It doesn't look down on lesser cars, although nearly every other car fades in its presence. And, the XJ-S driver doesn't get any resentment from other drivers - only Hi! signs, smiles and the right-of-way. Again, the enchantment of Browns Lane magic.

"All cars say something about their owners, but a Jaguar speaks volumes; its styling and performance respond in Shakespearean rhymes and couplets. And each verse on the XJ-S concludes with the same enthusiastic refrain; style and sport, class and performance."

At the end of June 1990, just in time to see Jaguar win at Le Mans for the seventh time, Sir John Egan left the Coventry company, handing over the reins to Ford veteran Bill Hayden CBE. Sir John (he was knighted in 1986), had turned around the fortunes of Jaguar - an achievement that must have seemed near impossible when he arrived at Browns Lane in spring 1980. Surprisingly, he moved away from the motor industry to take up a post with the British Airports Authority, but his contribution towards keeping the Jaguar legend alive will never be forgotten nor underestimated.

5

THE XJ-S IN COMPETITION

"The racing XJ-S had a cantankerous character. You had to command it and not let the car control you, but I have many happy memories ... People often ask me what was my most memorable race - I can't single one out, but I'll always remember standing on the podium, armed with a Union Jack and in Jaguar overalls, at Spa - a track which I love. It was a very emotional moment."- Win Percy, works driver.

The XJ-S's racing career started in the USA. The talented American racer, Bob Tullius, who'd previously gained some excellent results with Triumph TR4As and V12 E-types in the SCCA (Sports Car Club of America) series, put together an XJ-S racing programme under the Group 44 team banner.

Although the majority of SCCA events were supposedly amateur, be-cause of their impact on sales there was a lot at stake. Consequently, they attracted some very professional out-fits. Datsun was a perfect example, the company in the States establishing a competition department to supply uprated parts and offer advice to help as many Datsun drivers as possible.

Formed in 1965 and run by Tullius and Brian Fuerstenau, Group 44 was another highly-respected team, sup-porting BL products. From a Jaguar angle, Tullius started to prepare the V12 E-type for SCCA B-Production during 1974. Although they joined the season late, the Quaker State Oil-spon-sored machine immediately dominated the class. In the following year, Bob Tullius took the car to victory in the Road Atlanta Finals to claim the 1975 Championship, thus ending 17 years of Chevrolet Corvette supremacy.

In September 1975, the XJ-S made its debut in Frankfurt. Tullius could see the new car's potential, and, with the backing of Jaguar's New Jersey headquarters, plans were set in mo-tion to campaign the S in the 1976 season. By February that year, there was a new car in the Group 44 work-shops ...

The XJ-S in America

During the early part of 1976, Tullius was still campaigning (and very successfully) the V12 E-type, but by the end of the season was racing a specially-prepared XJ-S based on chassis 2W-51120. Finished in the traditional Group 44 white paintwork, the V12 coupé had a hard act to follow, as at that time, the team was winning

It was Group 44's 44th, and they won it with Quaker State.

The happiest fellow in the Georgia winner's cir-cle was Bob Tullius. He's been there before. But this time it's the 44th SCCA National Championship Race for his racing team, Group 44. And, sad to report, the last for his win-ning Jag. Happy retire-ment, you roaring V-12.

For the past 11 years Quaker State Racing Oil has been in the crankcase of every Group 44 entrant. At last count, that added up to about 300 wins and 14 national titles with Quaker State. Ask Bob Tullius why, and he'll tell you about Quaker State's super protection for the low-end bearings, its re-sistance to high-temper-ature foaming, and the way it keeps the machinery operating under tough racing conditions.

You don't go as fast as Bob Tullius, but you do drive a lot further before you change motor oil. And these days a car needs all the protection you can buy for it. Try Quaker State. We've been a winner for years.

Quaker State keeps cars running young.

Bob Tullius and the Group 44 team were already bringing Jaguar (and other BL-owned marques) a lot of good publicity. Bob was SCCA Champion with the E-type in 1975, and used the same car for part of 1976. By the end of that year, however, the Group 44 XJ-S had appeared on the scene.

an average of two out of every three races it entered.

The body was given flared wheelarches to allow for wider racing tyres, and, of course, a roll-cage was fitted, but apart from the necessary removal of trim and so on, there were surprisingly few changes. As Tullius said: "Probably the most surprising thing about the XJ-S is its aerodynamics. According to the sales brochures, Jaguar did a lot of wind-tunnel testing of the body design, but I figured half of that was BS and the other half was somebody dreaming. Well, let me tell you, now I'm a believer. To see if Jaguar's claims were anywhere near right we decided to keep this prototype XJ-S aerodynamically stock. You'll notice that we even use the production front spoiler. When I ran at Daytona I could effortlessly drive up and down the banking at 180mph."

The V12, developed under the watchful eye of Brian Fuerstenau, had proved exceptionally reliable in the E-type. The blueprinted engine was largely stock, except for the modified ignition, pistons, camshafts, valve springs and connecting rod bolts; on the induction side, an E-type manifold was employed to enable six twin-choke Webers to be fitted, fed by a 26.5 gallon fuel cell in the boot. A dry-sump system, designed and produced in-house at the Group 44 headquarters in Herndon, Virginia, looked after engine lubrication, while a standard radiator was used to keep it cool. A separate exhaust pipe for each bank exited just behind the two doors.

As for the transmission, this was a close-ratio, four-speed manual gearbox with a lightweight flywheel and Borg & Beck racing clutch. The differential was basically a stock item, except different ratios (ranging from 3.77:1 to 4.55:1) were employed, depending on the circuit. Both the gearbox and back axle were also given their own oil coolers.

Chassis work was mainly the responsibility of Lawton Fouchee, Group 44's Chief Mechanic, an ex-NASCAR man. The suspension was suitably stiffened all-round, with fabricated anti-roll bars and adjustable Koni shock absorbers added; at the rear, the standard trailing arms were replaced by new semi-trailing arms. The braking system was uprated using special Lockheed discs and eight-piston calipers up front, and the front brakes from an E-type at the rear; the single master cylinder was sourced from a Ford pick-up truck. Ten-inch wide Minilite magnesium alloy wheels were shod with Goodyear racing rubber, while powered steering was retained, albeit with less assistance.

Having not started at Watkins Glen, the XJ-S's race debut came in August at the Mosport TransAm event, where Tullius took pole position and claimed the fastest lap in Category I. Eventually, it finished fourth in Class, tenth overall. Next time out, at an SCCA event at Lime Rock, Tullius won after setting pole once again, and took a B-Production lap record in the process. At Indianapolis, the XJ-S didn't start (it had problems with its dry-

sump lubrication system), but Tullius had managed to amass enough points with his earlier performances in the E-type to qualify for the SCCA American Road Race of Champions.

At the Road Atlanta Finals in November, the 475bhp XJ-S duly qualified at the head of the field, but an uncharacteristic first-lap spin lost the battling Tullius a number of places before a fire once and for all put an end to his hopes for a sixth SCCA National title. There was also an aborted attempt at an IMSA event at Daytona - Tullius retired after lying fourth due to massive heat build-up in the cockpit caused by the location of the oil tank.

However, it was a very promising start, and for 1977, Tullius decided to forego the SCCA national events and

1976 SCCA

Watkins Glen	Tullius	DNS
Mosport	Tullius	4th
Lime Rock	Tullius	1st
Indianapolis	Tullius	DNS
Road Atlanta	Tullius	DNF

concentrate on Category I of the professional TransAm series (which was also run by the SCCA, incidentally).

1977 and 1978

The 1423kg racing S was now producing around 540bhp. As one would expect, this power gave the Group 44 XJ-S impressive performance. In 1977 trim, the car was capable of 0-60 in 5.0 seconds, and

Far left: The Group 44 XJ-S pictured in 1976.

Left: Impressive engine bay of the Group 44 XJ-S. Note the team's Triumph TR7 in the background.

1977 SCCA

Seattle	Tullius	1st
Westwood	Tullius	1st
Portland	Tullius	19th
Nelson Ledges	Tullius	8th
Watkins Glen	Tullius/Fuerstenau	4th
Hallett	Tullius	1st
Brainerd	Tullius	2nd
Mosport	Tullius/ Fuerstenau	1st
Road America	Tullius	DNF (race 1)
	Tullius	1st (race 2)
Saint Jovite	Tullius	3rd

Group 44's spaceframe car from 1981.

blasted through the standing-quarter in 13.6 seconds at a speed of 115mph.

The most impressive victory came at Mosport, where Tullius and Fuerstenau won the six-hour event outright. At the end of the season, having won five of the ten rounds in the TransAm series, Tullius was crowned Driver's Champion and Jaguar claimed the Manufacturer's title.

In the following year, Group 44 campaigned another XJ-S built especially for the 1978 TransAm season (although the old car did make two appearances to run alongside its newer stablemate). Lighter and with 560bhp

1978 SCCA

Sears Point	Tullius	9th
Westwood	Tullius	2nd
Portland	Tullius	3rd
Saint Jovite	Tullius	1st
Watkins Glen	Tullius/ Fuerstenau	1st
Mosport	Tullius	1st
Brainerd	Tullius	1st
Road America	Tullius	1st
Laguna Seca	Tullius/ Fuerstenau	1st 3rd
Mexico City	Tullius/ Fuerstenau	1st 8th

under the bonnet, it was to dominate the scene, with wins at Saint Jovite, Watkins Glen, Mosport, Brainerd, Road America, Laguna Seca and Mexico City. It was enough to once again secure both Driver's and Manufacturer's Championships.

Group 44's return

Sadly, after BL withdrew its sponsorship because of lack of funds, Group 44 didn't enter the S in 1979 or 1980, but returned to the tracks with the Big Cat in 1981. A change in SCCA rules meant that a spaceframe car with a 'silhouette' body could be entered. With the backing of the factory, a car was duly built.

This was a pure racer, weighing in at 1295kg, with a tubular frame, lightweight panels (many of which were aluminium), a custom-made suspension, quick-change differential, and a 525bhp, V12 engine moved further back in the engine bay to even up weight distribution. With Tullius at the wheel, the XJ-S won at Portland, Brainerd, and Mosport. Although Tullius couldn't add another TransAm title to the two he already held, he had won the most races that year and a second place overall in the championship was by no means a disgrace in a

1981 SCCA

Charlotte	Tullius	2nd
Portland	Tullius	1st
Lime Rock	Tullius	DNF
Road America	Tullius	DNF
Brainerd	Tullius	1st
Quebec	Tullius	DNF
Mosport	Tullius	1st
Laguna Seca	Tullius	5th
Sears Point	Tullius	DNF

brand new car.

The spaceframe car was to have one more outing, at the 1982 Daytona 24-hour event. For this race, Gordon Smiley was invited to share the driving with Group 44's Bob Tullius and Bill Adam. Clocked at a staggering 194mph, sadly, it was later plagued by gearbox problems. Finishing sixth in the GTX Class and 21st overall behind a gaggle of Porsches, it was not the team's most impressive result, but could have been better had it qualified for the GTO category. It should be noted that the S was due to compete in at least one more TransAm round, but the unfortunate death of Smiley whilst racing at Indianapolis put paid to the idea, and the car was retired.

As well as campaigning the XJ-S

Bob Tullius was entering the American-built XJR-5 in the IMSA GTP series. His Group 44 team entered two cars at Le Mans in 1984, signalling a long-awaited return to the Sarthe for the marque. Both cars retired, but they did acquire a great many fans in the meantime.

in SCCA and the lower class IMSA events, Group 44 went on to enter Jaguars in the highly-competitive IMSA GTP series. Egan was told about this, and after a certain amount of consideration, decided that Jaguar should back the project. The Lee Dykstra-designed XJR-5 made its debut at the Road America circuit in August 1982 and finished third. This was the start of the immensely successful XJR racing programme, both in the States and Europe, that eventually saw the Big Cats make a long-awaited return to Le Mans in 1984.

Racing in Europe

An interesting aside to the Leyland story was the racing programme. Ralph Broad of Broadspeed fame had managed to persuade the British Leyland directors to finance the running of a pair of XJ12Cs in Group Two of the European Touring Car Championship. At a time when Leyland needed all the good publicity it could get, it seemed like a decent proposal. In the event, the Broadspeed coupés just cost BL an awful amount of money (a six-figure sum has been quoted) and very little was gained from the project,

except perhaps a reputation for poor reliability.

After the 1976 and 1977 seasons, the cars were withdrawn, having not won a single race, despite a top line-up of drivers. Ironically, the XJ12C could have been a winner had Ralph Broad been supplied the parts he wanted and left alone to develop the cars the way he wanted. There was even talk at the works that the XJ-S could be used, but nothing came of it.

Andy Rouse, one of this country's finest drivers and race car engineers, was driving for Broadspeed at the time.

September 1977, and Andy Rouse takes the Broadspeed competition coupé into an early lead in the Tourist Trophy Race. Sadly, the car hit an oil patch and had to be withdrawn after looking set for victory. It seemed to sum up the Leyland campaign.

In a conversation with the author, he said: "There was a lot of politics. The crazy thing is that if the programme had been continued, they would definitely have won. The learning process was virtually over, and the competition had decreased. BMW's lightweight model, which was always the main threat, wasn't entered in the following year."

When John Egan arrived at Browns Lane, it was like a breath of fresh air. After it was announced that Group A cars (fairly similar to showroom models that are produced in quantities exceeding 5000 units per annum, but with a few suitable racing modifications) would replace Group 2 vehicles in the European Touring Car Championship for the 1982 season, he was approached by Tom Walkinshaw

regarding running the XJ-S in the ETCC.

Walkinshaw, born in Scotland in 1948, had started racing in Formula Ford in 1968, and within eight years was a works driver for BMW and on the verge of starting his own development engineering business. TWR - Tom Walkinshaw Racing - quickly became a force to be reckoned with. Egan could see the value in a racing programme, and duly promised the Scot technical co-operation and his full support if early results were encouraging - after all, the last thing Egan needed was another Broadspeed saga.

By September 1981, Walkinshaw and members of his Kidlington-based TWR team had drawn up guidelines for the way they wanted to tackle the project:

1. The car must be on the weight limit.
2. It must be easy to service.
3. Nothing outside the essentials to be included (applies particularly to cooling and lubrication systems).
4. No 'buy British' policy. Fit the best part for the job (e.g. BBS Mahle wheels, Bilstein shock absorbers, Recaro seat etc.).
5. Access to parts that need to be routinely serviced, replaced (choice of three differential ratios, for example) or repaired, under racing conditions or between races, must be good.
6. A minimum of 6mpg.
7. Engine to be developed for mpg and endurance rather than power, but it should produce no less than 380 bhp.

Fortunately, thanks to Australian John Goss wanting to use the S at Bathurst (he competed there from 1980

Tom Walkinshaw Racing (TWR) entered the XJ-S in the European Touring Car Championship. This is the 1982 car in full flight at a Donington test session.

The Walkinson/Nicholson XJ-S on the way to a fine victory in the 1982 Tourist Trophy race at Silverstone. It was followed home by Pierre Dieudonne and Peter Lovett in the TWR sister car.

Another shot of the 1982 TT-winning car, the nose dipping under hard braking.

to 1982, without success, sadly), the car was already homologated as a Group 1 machine, and most of the cars in this category were transferred to Group A following the changes announced for 1982.

Under Kevin Lee (TWR's Chief Mechanic) and Paul Davis (Team Manager), a three-man crew began work on the car in November. In the meantime, Alan Scott was assigned to the engine, while Eddie Hinckley looked after the chassis. By the early part of February, Walkinshaw was testing the XJ-S at Gaydon.

In body terms, Group A regulations stated that very little could be changed; of course, it was strengthened by seam welding the shell and adding a roll cage, but these modifications, along with stripping the interior,

were purely for the sake of safety and standard practice for competition cars. The uprated engine (said to be developing 400bhp at 7000rpm) was mated to a four-speed manual transmission, and given additional cooling to allow them to work at speeds of up to 175mph. A stiffer suspension and racing tyres completed the 1400kg package. The scene was set ...

1982

Although the TWR XJ-S's debut at Monza ended in retirement, it did lead the race by a fair margin until an unfortunate coming together, and Walkinshaw's professionalism impressed everybody. *MotorSport* noted: "Just a few months after the project was first seriously voiced by TWR to sponsors Motul, the French oil

concern, the car appeared at Monza for the March opening round of the 1982 European Touring Car Championship. It was immediately obvious that Tom Walkinshaw, who was sharing the driving with 'Chuck Nicholson' [actually, his real name was Charles Nickerson], had tackled the problems of racing a Jaguar with a professional acumen that was totally absent in the original over-complex Broadspeed XJ Coupé project."

Walkinshaw won a non-championship event at Zolder in April, but it was victory in the ETCC that was so crucial. At least this win proved the car was not only fast but also reliable enough to bring the Kidlington outfit the results it wanted in time.

Eventually, a second car joined the team and victory went to the TWR

Walkinshaw at the end of a gruelling Tourist Trophy. Nowadays, he has put away his racing overalls in favour of managing full-time. Among other things, at the time of writing he ran the Arrows F1 team.

1982 ETCC

Monza, ITA	Walkinshaw/ Nicholson	DNF
Vallelunga, ITA	Walkinshaw/ Nicholson	3rd
Donington, GB	Walkinshaw/ Nicholson	DNF
Mugello, ITA	Walkinshaw/ Dieudonne	DNF
Brno, CZH	Walkinshaw/ Nicholson	1st
Zeltweg, OST	Walkinshaw	2nd
Nurbürgring, GER	Walkinshaw/ Nicholson	1st
Spa, BEL	Walkinshaw/ Nicholson/Percy	DNF
	Dieudonne/ Allam/Lovett	DNF
Silverstone, GB	Walkinshaw/ Nicholson	1st
	Dieudonne/ Lovett	2nd
Zolder, BEL	Walkinshaw/ Nicholson	1st
	Dieudonne/ Allam	2nd

drivers at Brno and the Nürburgring. It then came as no real surprise when TWR XJ-Ss achieved first and second places in the Tourist Trophy Race at Silverstone; it was Jaguar's first TT victory for 19 years. With a win at Zolder as well, TWR had won four rounds of the ETCC in 1982, and thus deservedly secured the backing of the factory for the following season.

Jim Randle, Jaguar's engineering boss, was excited at the prospect of the marque's return to the race tracks. He said: "First of all, racing focuses inter-est, within a company as well as outside it. It is good for morale and for our engineering disciplines. One of the great features of Group A is that the exterior shape of the cars must remain virtually unaltered, so they don't look like freaks. Another feature of the ETCC that appeals to us is that the races aren't just flat-out sprints; they are at least 500km long - and there are 12 of them in the year.

"When we gave up racing, all those years ago, we never lost the dream of returning one day; but times change, and no matter how interested you are in the sport you can't begin 'cold' and expect to be winning right away. We decided we must make use of the best possible experience at our disposal - and no British-based team had more to offer than TWR."

1983

With support from the works secured, the TWR cars now ran in a white with green livery instead of the predominantly black colour scheme seen the previous year. A few mechanical changes were also witnessed, with the adoption of outboard rear brakes and a new fabricated rear suspension. The braking system as a whole was uprated once again, and 17 inch Speedline alloy wheels (some 13 inches wide!) were employed in place of the former

The TWR XJ-S sporting its new 1983 livery.

1983 ETCC

Monza, ITA	Walkinshaw/ Nicholson	2nd
	Dieudonne/ Calderari	DNF
Vallelunga ITA	Dieudonne/ Nicholson/Walkinshaw	3rd
	Walkinshaw/ Calderari	DNF
Donington, GB	Brundle/Fitzpatrick/ Calderari	1st
	Walkinshaw/ Nicholson	5th
Pergusa, ITA	Walkinshaw/ Nicholson	1st
	Calderari/ Dieudonne	DNF
Mugello, ITA	Walkinshaw/ Fitzpatrick	3rd
	Calderari/Nicholson	3rd
Brno, CZH	Walkinshaw/ Nicholson	1st
	Dieudonne/Calderari	6th
Zeltweg, OST	Walkinshaw/ Brundle	1st
	Dieudonne/ Calderari	2nd
Nurbürgring GER	Walkinshaw/ Nicholson	DNF
	Dieudonne/ Calderari/ Walkinshaw	DNF
Salzburgring OST	Walkinshaw/ Nicholson	1st
Spa, BEL	Walkinshaw/ Dieudonne	DNF
	Percy/Calderari/ Brundle	DNF
Silverstone, GB	Walkinshaw/ Dieudonne	9th
	Nicholson/ Dieudonne	DNF
Zolder, BEL	Walkinshaw/ Brundle/Percy	8th
	Percy/Dieudonne/ Calderari	DNF

BBS items. Initially, the four-speed Jaguar gearbox was retained, but a Getrag five-speed unit took its place as the season progressed.

At the first round, the two TWR XJ-Ss qualified at the front of the grid, and were genuinely unlucky not to win the race (a loose bonnet called for an unscheduled pitstop). There was a third place for Jaguar at Vallelunga, but at Donington, Martin Brundle (who had just joined TWR, along with John Fitzpatrick) managed to take the flag for the Kidlington team.

Pergusa was next, and there was another victory, this time for Tom Walkinshaw and Chuck Nicholson. A third place followed at Mugello, and Walkinshaw's win at Brno gave him the lead in the Driver's Championship, a position he retained after a TWR one-two at Zeltweg.

Both cars retired at the Nürburgring (it was to be the last time the old circuit was used for an FIA-sanctioned event), but there was another win at the Salzburgring. The team had a disasterous race at Spa, and Silverstone wasn't much better. At the final round at Zolder, on 25 September, Win Percy (who had been BTCC Champion in a TWR-prepared Mazda RX7 in 1980 and 1981) led the race for TWR, but had to retire; the best placed Jaguar finished eighth.

By the end of the season, the XJ-S had taken no less than five wins in the ETCC, but the BMWs claimed six. Walkinshaw finished second in the Driver's Championship to Dieter Quester, while Jaguar was second in the Manufacturer's Championship.

1984

Now finished in a more suitable British Racing Green, TWR announced it would be running a three-car team for 1984. Walkinshaw would be joined by Percy, Brundle, Nicholson, Hans Heyer and Enzo Calderari. David Sears (son of Jack Sears - a respected racer during the 1950s and 60s), joined the team later in the year.

Walkinshaw and Heyer won in the first race of the season at Monza, and this was followed by a third at Vallelunga; Percy and Nicholson then won at Donington. At Pergusa, the XJ-S recorded a magnificent one-two-three finish, a feat that was repeated at Brno.

The TWR cars took the first two

Emerging from the mist, the car shared by Martin Brundle, John Fitzpatrick and Enzo Calderari, which took victory at a rain-soaked Donington in 1983.

At Spa in 1983, Walkinshaw (who had teamed up with Dieudonne on this occasion) had trouble with the propshaft; then the differential failed. For this event, US-type headlights were used.

Multiple BTCC Champion and all-round good chap, Win Percy (left) joined the TWR Jaguar team full-time for 1984, as did ex-BMW man Hans Heyer, seen here on the right.

The grid at Donington, April 1984.

1984 ETCC		
Monza, ITA	Walkinshaw/Heyer	1st
	Brundle/Calderari	13th
	Percy/Nicholson	DNF
Vallelunga, ITA	Walkinshaw/Heyer	3rd
	Calderari/Nicholson	8th
	Percy/Schlesser	DNF
Donington, GB	Percy/Nicholson	1st
	Brundle/Schlesser	5th
	Walkinshaw/Heyer	9th
Pergusa, ITA	Calderari/Brundle	1st
	Walkinshaw/Heyer	2nd
	Percy/Nicholson	3rd
Brno, CZH	Walkinshaw/Heyer	1st
	Percy/Nicholson	2nd
	Calderari/Sears	3rd
Österreichring OST	Walkinshaw/Heyer	1st
	Percy/Nicholson	2nd
	Calderari/Sears	DNF
Salzbürgring OST	Percy/Nicholson	1st
	Calderari/Sears	2nd
	Walkinshaw/Heyer	DNF
Nurbürgring GER	Walkinshaw/Nicholson/Heyer	5th
	Percy/Calderari	DNF
	Heyer/Sears	DNF
Spa, BEL	Walkinshaw/Heyer/Percy	1st
	Calderari/Sears/Pilette	DNF
Silverston, GB	Calderari/Sears	2nd
	Walkinshaw/Heyer	DNF
	Percy/Nicholson	DNF
Zolder, BEL	Walkinshaw/Heyer	3rd
	Calderari/Percy	4th
	Percy/Nicholson	DNF
Mugello, ITA	Brundle/Calderari/Sears	5th
	Walkinshaw/Heyer	DNF
	Percy/Nicholson	DNF

places at the Osterreichring; there was another one-two at the Salzburgring, but fifth was the best anyone could do at the Nürburgring. At Spa, Walkinshaw decided to enter two cars for the 24 hour event - one retired, but the Walkinshaw/Percy/Heyer machine took the chequered flag literally miles ahead of the competition. It was the first Jaguar victory in an International 24 hour race for 27 years!

A second at Silverstone was fol-lowed by a third at Zolder. Brundle was the only Jaguar to complete the course at Mugello (in fifth), but by the end of the 1984 season, the TWR XJ-S had captured the ETCC title with seven victories. Tom Walkinshaw took the Driver's Championship, with Heyer in second place.

Away from the ETCC, John Goss asked Tom Walkinshaw if he would partner him at Bathurst - Australia's classic race. Sadly, the car retired on

Win Percy and Chuck Nicholson took victory at Donington in 1984; their team-mates finished fifth and ninth.

Jaguar at the first ETCC race on the Neue Nürburgring in 1984. Sadly, having started one-two-three on the grid, it turned out not to be a good event for the TWR team.

The European Touring Car Championship was Tom Walkinshaw's in 1984, after a dominating performance by the Jaguars. This action shot from the pits was taken at the round held at Spa.

the start line, but in November, TWR entered two cars in the Touring Car race of the Macau Grand Prix. Sponsored by John Player Special, the XJ-Ss displayed the traditional black and gold JPS colours, and went on to dominate the event. Walkinshaw finished first, with Heyer second - it was the perfect end to a perfect year.

1985 and beyond

There was no involvement in the ETCC for 1985. Although it was a shame not to see the Big Cats in action, it was perhaps better that they should be remembered as winners. However, Walkinshaw returned to Bathurst for the James Hardie 1000 - which was

now being run under Group A regulations - with a proper team of three cars. Armin Hahne and John Goss took the flag ahead of a Schnitzer-prepared BMW 635CSi, with the Walkinshaw/Percy XJ-S in third.

Walkinshaw took two cars to the Fuji circuit in Japan at the end of 1986, but both retired. As something of a swansong (its homologation papers were about to run out), TWR entered two events in New Zealand in the opening months of 1987, one in Wellington, and one in Pukekohe in February. Both cars failed to finish in the Wellington four hour event but, finishing just behind a Holden Commodore, Win Percy and Armin Hahne took their

XJ-S to second place in Pukekohe.

In the meantime, TWR had teamed up with Jaguar on a different project, this time on the XJR racing programme. A number of the XJ-S ETCC campaign members moved over to the TWR Group C team, including Martin Brundle (one of the drivers who took the XJR-12 to victory at Le Mans in 1990) and Win Percy.

Designed by Tony Southgate, the XJR-6 was the first of the TWR line, making its debut at Mosport in August 1985; Brundle took third place after making the early pace. Walkinshaw's commitment gave Jaguar the World Sportscar Prototype Championship title in 1987, 1988 and 1991, and wins

Action from the last ETCC round of 1984, by which time, the championship had long since been sealed. Only seven Group A cars were built, but, between them, they scored an enviable amount of race wins. They were extremely quick, with 0-100mph coming up in under ten seconds, and the capability to lap the banked circuit at Millbrook at an average speed of 176mph!

The TWR XJR-6 entered the WSPC scene during the latter half of the 1985 season. It made its debut at Mosport, Canada, and paved the way for a whole series of XJR Group C and IMSA GTP cars.

The Jaguar XJR-9LM on its way to victory in the 1988 Le Mans 24-hour race. Without the success of the Group A XJ-Ss, it's doubtful whether the WSPC programme would have ever got off the ground.

at Le Mans in 1988 and 1990. By this time, TWR had successfully taken Group 44's place as the works-backed team in the IMSA series, winning a number of races.

Looking back, the TWR XJ-S racing programme had been extremely successful. Seven cars had been built in all, claiming between them 16 victories in the ETCC. Other wins at Macau and Bathurst brought excellent publicity in the Far East and the Antipodes, while Group 44's previous successes in America assured the S legendary status in racing circles around the world.

6

REBIRTH OF THE GT LEGEND

At the 1990 Motor Show, held at the NEC, Jaguar displayed the 1991 model year range. A catalytic converter now came as standard on the V12 models (power was reduced to 273bhp), but the quoted top speed of 148mph for the

America's 1991 model year XJ-S, seen here in Classic Collection guise (note the discreet 'Classic' badge on the right-hand side of the bootlid and gold-plated 'growler' on the bonnet), with a couple of other Jaguar classics - the C and D-types.

THE JAGUAR XJ-S

The Classic Collection cars came with unique colour and trim combinations, the leather (usually magnolia or doeskin) also extending to the matching gearknob; the leatherbound steering wheel was finished in a charcoal colour.

Fascia of the American 1991 MY Classic Collection XJ-S. The S had certainly come a long way in its 15 year production run, but a major change was just around the corner.

The cockpit of the XJ-S invites you to take command of all the power and performance that this V-12 Jaguar has to offer. A full array of analog gauges monitor speed as well as engine performance and condition. The tilt steering wheel, which also houses an air bag, offers six different positions to suit the driver. The side mirrors adjust electrically and heat in harsh weather to remove frost and moisture buildup. A switch on the large center console initiates cruise control for long trips.

Above the center console, a seven-function trip computer displays useful travel information, such as average and instant fuel economy, time of day, elapsed time, total fuel consumed, average speed and total distance traveled.

As elegant as it is purposeful, the S-type cockpit also offers a wealth of convenience, comfort and entertainment features. A computerized climate control system regulates heating or cooling according to the selected temperature. Jaguar's 80-watt stereo sound system plays through four acoustically matched speakers, providing rich, resonant sound for you and your passengers.

Right: The new XJ-S. This is the four-litre coupé, as seen in the first UK catalogue. The tail-end treatment and revised rear quarter-light styling were the most noticeable changes.

riod, but at least there was a new XJ-S ...

A new XJS

No, the above sub-title is not a mistake! From now on, instead of being designated the XJ-S, Jaguar's sporting model would drop the hyphen it had carried since launch to become the XJS. Although the S was still attracting reasonable sales for such an old concept, as Jim Randle pointed out "It was withering on the vine." Two choices were left open to Jaguar: either let the S fade away into the history books or revive it. Fortunately, the latter course of action was chosen.

With the same Project Team that brought the Convertible to life, before reaching the market the new model underwent an extensive development programme, taking in over one million miles of durability testing in extreme conditions, from freezing Ontario to the deserts of Arizona.

Announced on 1 May 1991, styling changes were carried out by Geoff Lawson (Jaguar's Chief Stylist since 1984). Of the 490 panels, 180 were either new or modified, but it was a very subtle facelift, the main identifying features being new rear side windows and rear lights.

Tooling again came from Karmann of Germany, but panels were sourced from Venture Pressings - a joint project between Jaguar and GKN. Bill Hayden was quick to point out the benefits in quality control: "We are taking our first panels from the Venture Pressings operation, which will give us complete control over body supply. We have also invested in new manufacturing facili-

closed XJ-S was still very impressive. At the same time, a limited edition 'Le Mans V12' was announced to celebrate the company's victory at the French circuit that year.

The 'Le Mans V12' featured "a sports suspension, special colour range, 16 inch forged lattice alloy wheels with gold centre spokes, unique badging and grille, colour-keyed door mirrors and quad headlamps." As for the interior, there was a "revised seat style in cream Autolux leather, Wilton carpeting, sapwood veneer on the centre console and door top rolls, a four-spoke leather-trimmed steering wheel, and limited edition numbering and badging on the sill treadplates. The leatherbound handbook wallet is embossed with the 'Le Mans V12' logo, as are the headrests." Only 280 were built.

JaguarSport's 318bhp XJR-S continued pretty much unchanged as attention shifted to the supercharged XJ40 - the XJR. Tom Walkinshaw, JaguarSport's MD, said: "The response to our recent European six-litre XJR-S launch has been tremendous and we intend to follow up this success with the launch of the XJR four-litre saloon

in European markets in October. Our next target is the United States."

A total of 48,138 Jaguar and Daimler models were built in 1989, but the recession soon had an effect on the executive market; by 1992, production had dropped to just 20,601 units, as even American sales tailed off sharply. With the recession biting hard, the early 1990s were not a good time for luxury car makers. Indeed, despite recent good financial results, production was cut back in 1991, with the company posting a £226m loss.

Under Ford, however, with its massive resources, Jaguar at last had the finances in place to introduce new models, and Bill Hayden was just as determined to make the Coventry firm succeed as Sir John Egan had been.

Hayden, in keeping with his reputation for giving companies the short, sharp, shock treatment, said on his arrival: "Our objective in all this is quite simple. It is to produce better and better quality products here at Jaguar. Our commitment to that objective will be rigorous and unrelenting." Sadly, the infamous F-type was cancelled after a very protracted development pe-

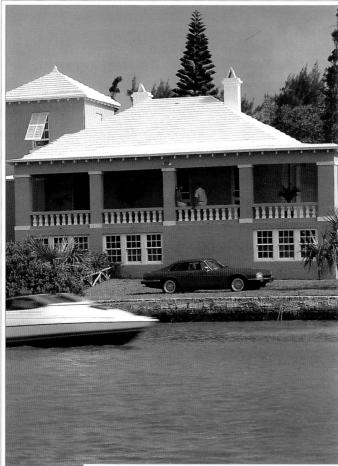

Jaguar XJS 4.0 Coupé

*I*t's performance dramatically boosted by a new 4.0 litre 24 valve engine — torque is increased by some 12% over the previous 3.6 litre unit — the new Jaguar XJS 4.0 Coupé brings fresh exhilaration to 6-cylinder XJS motoring.

Power is transmitted through a 5-speed manual transmission system. (A 4-speed automatic gearbox is available at extra cost with a switchable choice of 'normal' or 'sport' mode) 60 mph is attainable from rest in 7.5 seconds (0-100 km/h in 7.9 secs) with a maximum speed - where legislation and conditions allow - of 141mph (228 km/h).

Styling is dramatic, the new matt black radiator grille and colour keyed wiper grille emphasising the long bonnet. Restyled frameless tinted windows lead the eye to the redesigned rear wing panels, contoured boot and neutral density rear lamps.

Within this 2 + 2 coupé, colour co-ordination is comprehensive. New style sports seats are trimmed in leather and tweed, enhanced by box pleating and double needle stitching. Leather trim is an extra cost option. Each front seat is electrically adjustable for recline angle and fore and aft movement by means of door-mounted switches which, for added convenience, can be operated from outside the car once either door is opened - even with the ignition switched off.

An electronic memory (an extra cost option on the XJS 4.0 Coupé) returns the driver's seat - and the electrically adjustable exterior mirrors - to one of two pre-set positions at the touch of a button. Electric heating and silently adjustable lumbar support are further front seat luxury options.

Travelling pleasure is enhanced by full air conditioning and a sophisticated audio system (described on page 31).

Driving aids abound. New analogue instruments are framed by burr walnut-grain veneers. A row of 'secret till lit' advisory and warning lamps includes a reminder of the currently selected automatic gearbox mode. Cruise control and a heated headlamp powerwash system are extra cost options. Equipped with anti-lock braking and a built-in sports suspension system*, the new Jaguar 4.0 Coupé is a pleasure to own and drive.

* *The sports suspension system fitted to the Jaguar 4.0 Coupé is similar in design and function to the optional Sports Suspension System described on page 32.*

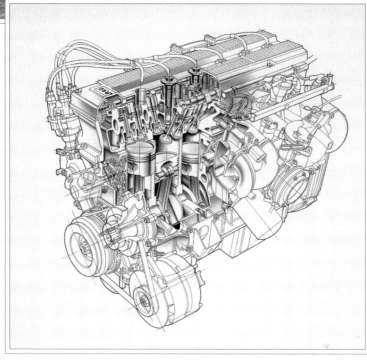

A cutaway drawing of the four-litre AJ6 power unit.

ties so that we can achieve improved fit and finish, as well as more efficient assembly." Indeed, almost £4 million was spent on the Castle Bromwich plant, equipping it with the latest computer-controlled systems, and a further million was spent on Track 3 at Browns Lane.

There was also a four-litre engine to replace the old 3.6 litre six. The four-litre version of the AJ6 engine was launched in the XJ40 saloon at the 1989 Frankfurt Show. There was little change to the unit for its XJS application. It retained the same all-aluminium, lightweight, closed-deck structure, four-valve, pent-roofed combustion chamber, and duplex chain to drive the direct-acting valvegear.

The increased displacement of 3980cc was obtained by lengthening the stroke from 92 to 102mm. Although

Jaguar XJS V12 Coupé

Dominating the impressive specification of the new Jaguar XJS V12 Coupé is its race-bred 12 cylinder engine - source of seemingly endless reserves of smooth, near-silent power. One of the world's most potent production engines, this classic 5.3 litre unit is as well-mannered in the rush hour crawl as it is on the autobahn, responding to every command with calm flexibility. Digital ignition is complemented by a new fuel control system that combines efficient all-weather starting and engine warm up with diagnostic fault retrieval.

Performance is impressive. 60 mph can be reached from a standstill in 7.8 seconds (0-100 km/h in 8.2 secs) and - legislation and conditions allowing - a maximum speed of 147 mph (237 km/h) is attainable.

Power is transmitted through a sophisticated automatic gearbox teamed with cruise control. Electronic anti-lock braking, limited slip differential, power steering and fully independent suspension ensure that the V12 power output is matched by superb handling qualities.

The substantially re-styled interior is distinguished by several unique luxury features.

One of these is the lavish use of supple leather. As many as three selected hides per car are used for the new box-pleated seat facings, centre console, central armrest (with concealed cubby box) and steering wheel pad. Each car-set of hides is carefully matched for colour and grain before being cut and sewn by skilled hands. The extensively colour-keyed decor is enriched by handsome burr walnut veneers, highlighted with matchwood inlays and hand-polished by craftsmen to a rich lustre.

Both front seats are electrically adjustable for recline and fore-and-aft movement (as described on page 29). The memory control is fitted as standard to the V12 Coupé.

The audio system combines electronic sophistication with refreshingly simple operation. The system is wired for an optional compact disc (CD) autochanger which is housed in the boot and which holds up to six discs, individually selectable from the driver's control panel. RDS (Radio Data System) automatically retunes the radio to the strongest signal for each station as your journey progresses and is capable of interrupting tape or CD playback to relay traffic news from local FM stations.

Luxurious as it is, the new Jaguar XJS V12 Coupé is first and foremost a driver's car. Every control is at the driver's fingertips, including heated screenwashers and headlamp powerwash, front foglamps, central locking (including the boot lid and fuel filler flap), and electrical adjustment of the heated door mirrors.

Seated before the leatherclad steering wheel, the Jaguar V12 Coupé driver is in command of every situation.

A dynamic invitation

Jaguar XJS V12 Convertible

Designed and built in Britain, the powerful and versatile Jaguar XJS V12 Convertible is well suited to the national climate, where sudden showers are apt to dampen wind-in-the-hair enjoyment. At the touch of a button, the tailored hood and pillarless rear windows close, affording you and your passenger the snug comfort of a luxurious fixed-head coupé. With the hood up or down, the Jaguar XJS V12 Convertible offers keen drivers a plenitude of delights, beginning with 5.3 litres of high-powered authority. Virtually silent in action, the hand-assembled V12 engine is further refined by a new fuel control system.

Silk smooth automatic transmission, power assisted steering, electronic anti-lock brakes, new analogue instruments, new electronic cruise control and trip computer, comprehensive central locking, air conditioning and an exclusive RDS audio system with CD compatability - this is a recipe for total satisfaction.

Test drive the Jaguar XJS V12 Coupé and you will surely be impressed by the positive handling and comfortable ride. So it may come as a surprise to learn that there is the extra-cost option of a Sports Suspension System which offers even tauter handling, with less body roll on fast bends and reduced float and pitch.

The system includes a number of critical changes to the standard suspension and steering settings. Spring rates are increased at both the front and rear, a larger front anti-roll bar is fitted and dampers are retuned.

Power steering assistance is slightly reduced, providing greater 'feel' and emphasising the inherent precision and keen response of the rack and pinion system.

The new XJS V12 Coupé already exhibits a strong sense of purpose and this is accentuated by fitment of the Sports Suspension System, which lowers the ride height to give tauter handling and creates an even more dynamic stance. This visual theme is extended by the fitment of 7" x 16" (178 x 407 mm) forged alloy lattice wheels, which are also available as a separate extra-cost option*. Lighter and brighter than the standard 15" (381 mm) cast alloy wheels - and looking remarkably like traditional 'wire' wheels - they are shod with lower profile P600 225/55 ZR16 tyres.

Whilst engendering tauter and even more responsive handling, the Sports Suspension System is entirely compatible with the high standards of ride refinement for which Jaguar XJS V12 Coupés have won a world of respect.

* These 16" (407 mm) forged lattice alloy wheels with 225/55 ZR16 low profile tyres are also available as an extra-cost option for Jaguar XJS 4.0 Coupés.

A car for all seasons

It's easy to make the most of fine weather when you own a Jaguar XJS Convertible.

Opening the power-operated hood takes a mere twelve seconds. With the gear selector in 'Park', simply release the lockdown catches above your head and press a switch on the centre console. The hood glides open, stowing itself neatly behind you (a padded hood cover is provided), and the rear quarter windows open simultaneously.

Lower the front electric windows and you are ready to enjoy the rare delights of a fully open car.

Closing the hood and rear windows is just as quick and easy. If rain remains a threat, you can drive with the hood closed, whilst retaining glassless and pillarless vistas on either side.

Cold weather presents no problems. The hood - now tailored in a choice of four colours - is fully lined and insulated to keep draughts out. The glass rear window is tinted and electrically heated. Both seats are electrically heated (with adjustable lumbar support) and feature - as in the V12 Coupé - electrically adjustable fore, aft and recline movement with memory control*.

Behind the seats is a full-width lockable storage box.

As in the V12 Coupé, a wealth of driving aids is matched by the beauty of hand crafted leather upholstery and inlaid burr walnut veneers.

Sleek looks, V12 power, excellent handling, unashamed luxury and a weatherproof hood that opens or closes in a matter of seconds; come rain or shine the new Jaguar XJS V12 Convertible makes every journey a luxurious pleasure.

* Electric seat adjustment is described on page 29

---

there was a very marginal increase in power (to 223bhp at 4750rpm), the engine featured redesigned pistons and a forged steel crankshaft (instead of cast-iron) to improve engine refinement, while a revised cam profile with increased lift and reduced overlap improved both low-end torque and idle stability.

The Lucas 15CU engine management system gave enhanced starting and idle speed control, as well as the ability to communicate electronically with the control unit on the new ZF four-speed, dual-mode, automatic transmission, thus improving shift quality. With a compression ratio of 9.5:1, maximum torque output was quoted as 278lbft at 3650rpm.

However, the engine and transmission had been around for a while, so not surprisingly it was the attractive styling that made the headlines. Some critics were very impressed by the way in which the lines had been modernised whilst carefully preserving the flavour of the original.

As Mike Dale, the President of Jaguar Cars Inc., said: "Our customers don't like surprises. The XJ-S is a classic design, like the Range Rover or the Porsche 911." After all, if a clinic for Jaguar owners gives the styling of the HE a rating of 98%, it is crazy to make dramatic changes. For some, though, the facelift was a little too subtle.

As William Kimberley wrote in *MotorSport* : "For all the £50 million spent on the new model and the 40% new body panels, only the keen enthusiast will be able to spot the differences between the new model and its predecessors, especially when looking from the front. It is true that the radiator grille, headlamps and windscreen are different, but you have to be an expert to tell. It is also true that the coupé, while retaining the famous flying buttresses, has a greater glass area, but this turns out to be a cosmetic job, the plastic panelling of the rear window having now become blackened glass with an imprinted XJS logo which has done nothing to improve vision for the driver. From the rear, though, the new bootlid, extra chrome and full width lighting are the giveaway that this is the new model, together with the new fuel filler flap, the flared sills and, as an option, the 16 inch forged alloy lattice wheels.

"Once inside the car, though, both driver and passenger will notice a difference. Everything is much more attractively laid out, aided by the substitution of the burr elm wood cappings by burr walnut. The switchgear is new, the air conditioning uprated and and there is an improved stereo system with CD as an option. Most important of all, however, are the improved two-position memory seats which are also linked to the door mirrors. The interior is now more befitting of a Grand Tourer than before.

"While a decision still has to be made about whether the Convertible will be fitted with the four-litre engine, its installation with the catalytic converter in the coupé has enabled the six-cylinder XJS to be sold in Austria, Switzerland, Sweden, Canada, Australia, South Africa and Japan for the first time, which should ensure that the sales figures keep above the 10,000 mark, despite the industry-wide slump in home sales."

While the four-litre Convertible was not to join the line-up for a full year, the 5.3 litre V12 models were retained in the range, but naturally with the new body styling and interior appointments. The engine was now producing 280bhp, as a low-loss catalytic converter was fitted as standard. However, the HE designation had been lost, so that the two twelve-cylinder cars were known as the XJS V12 and the XJS V12 Convertible respectively. A Sports Suspension Package was available on the V12 coupé but, for the time being at least, was not offered on the open car.

Profile shot of the new XJS, this being the four-litre coupé.

Although the new XJS didn't look that much different to its predecessor from the front, the differences were obvious in the tail treatment.

In the UK, the four-litre coupé was priced at £33,400 at the time of its launch (automatic transmission added a hefty £1380 to this), whilst the V12 models started at £43,500 for the coupé, rising to £50,600 for the Convertible in standard trim.

Jaguar's Chairman, Bill Hayden, said at the time: "One of the things that sets Jaguar apart from any other motor manufacturer is the pride and the passion that the name still evokes. Over the last two months I have witnessed this first hand.

"I am delighted with the new range. It is the best XJS yet. We have freshened the car's appearance without losing its distinctive styling. With the new four-litre engine, improved V12 and the interior facelift, we have an outstandingly competitive XJS to offer sports car buyers here and abroad."

Autocar & Motor carried out a test on an automatic four-litre coupé, and was impressed with the refinement, comfort, taut handling and much improved build and finish, but disappointed by performance, the meagre space and styling. Obviously, the magazine was expecting more from Jaguar's stylists than "a competent but rather

Jaguar's new Chairman, Nick Scheele, with the award-winning XJ40 saloon, this being the 1993 model year XJ6 3.2. A two-door version of the XJ40 was considered, and a prototype built, but the idea was not developed further.

perfunctory nose and tail job."

However, performance figures were quite interesting. With a 3.54:1 final drive and gearbox ratios of 2.48, 1.48, 1.00 and 0.73:1, the maxima through the gears was recorded as 50, 83, 124 and 137mph. 0-60 came up in 8.7 seconds, and the standing-quarter covered in 16.7 seconds at a speed of 87mph. Overall fuel consumption averaged out at 18.8mpg; not bad for a 1612kg car that has been driven hard.

Testing the same vehicle, *Performance Car* noted: "On the open road the XJS feels quicker than its on-paper figures suggest, due in part to the responsiveness of the dual-mode ZF automatic gearbox. In 'Sport', the transmission changes down more readily on part-throttle acceleration and holds on to the revs to the red line at 6250rpm. Although kickdown is readily achieved when overtaking, manual operation of the 'box gives a quick and smooth response, for example when braking down for a corner. 'Normal' mode allows for a less frantic approach to acceleration and, consequently, marginally better fuel consumption.

"Power delivery is clean and unfussed, if not quite as smooth as the best from BMW and Mercedes, and the straight-six emits a pleasingly sporting note under full throttle. It is, however, easy to conceal engine noise through careful acceleration, and high speed motorway cruising is a serene affair, the engine remaining virtually inaudible at most speeds.

"Jaguar has taken enormous steps forward in terms of quality control with the new XJS. For the most part, the car is now well up to its main rivals from Germany and Japan, and can even show some of them a thing or two, particularly in ride and handling and sheer individuality.

"The new four-litre engine provides plenty of performance, together with decent fuel economy, and delivers it in a characterful way that will please

The UK specification four-litre Convertible at the time of launch.

the sporting driver, and yet remains unobtrusive when requested."

Reviewing the latest V12 Convertible, *What Car?* gave the new model a lowly two out of five for accommodation, four out of five for the behind-the-wheel category, performance and economy, quality and equipment, and service and costs. It gained full marks in the handling and ride department and, more importantly, in the overall verdict on the machine.

Nick Scheele took over as Jaguar Chairman in March 1992, having previously been in charge of Ford's operations in Mexico. A very likeable character, Scheele had been groomed for his position at Browns Lane for some time.

Just two months after Scheele's arrival, the four-litre Convertible made its debut. Priced at £39,900 (automatic transmission added £1370 to this figure), it was an impressive piece of machinery. "Jaguar has devised a cunningly simple solution to the quiver and shake owners have had to endure up to now," declared *Car*. "By no means did it ruin the XJS Convertible experience, but it did rather undermine the tranquillity ... Well, the good news is that it's almost completely fixed, and the solution lies in a couple of lengths of stainless steel tube which are knitted together as an X, and bolted across the bottom of the engine bay. Crude, perhaps, but it sure does quell the shiver and quake. And there's more. The ZF four-speed auto transmission and its propshaft have been modified to improve shift quality, and there's now an option on a driver's side airbag at £700."

By the end of the year, a new four-spoke steering wheel was fitted across the XJS range with a driver's side airbag as standard, the four-litre coupés could be specified with a 'Touring' suspension package, the soft-top models benefited from the addition of a rear underfloor strut assembly ("to further minimise body shake"), there was a new velour trim for the six-cylinder cars, and the improved automatic transmission featured a first gear inhibit.

MotorSport pointed out: "The performance of the uprated chassis is most impressive, as is the way that one can glide to 100mph (at just 3500rpm) with a complete lack of fuss. The ride and handling compromise simply eliminates bad manners and must make rival development engineers despair that they cannot do better with more modern running gear."

There was also a new 'Insignia' service offered for customers who wanted to personalise their cars. Using many of the craftsmen from the old Daimler Limousine shop, the newly-formed Special Vehicle Operations Department was, in the words of Roger Putman, Jaguar's Director of Sales & Marketing, established to give "the customer tremendous choice and flexibility in selecting from a new range of exterior colours and all-leather interior trims. Customers can match these aspects of the car to personal taste."

US update

In America, the new XJS arrived in autumn 1991, just in time for the 1992 model year. At last, the twin-headlights that had been a feature of all XJ-Ss headed for the States were dropped when the 1992 S was launched. In the meantime, the old HE-shape models continued, priced at $55,905 and $67,565 respectively for the coupé and convertible. However, there were few complaints. As *Car & Driver* stated: "The Jaguar 5.3 litre V12 engine has been graced with a new Lucas 26CU fuel system to go with its Marelli digital-ignition system, and it remains as satisfyingly smooth and silky as ever, even when compared with newer V12s from BMW and Mercedes."

Sales were still falling in America due to the financial climate (from a peak of 24,464 cars - including saloons - in 1986 to only 8681 in 1992, of which just 2058 were XJSs), but at last, Egan's drive for quality was paying dividends, putting Jaguar in the top ten of the 1992 JD Power & Associates Customer Satisfaction Index, finishing ahead of BMW, Porsche, Volvo and Saab.

For 1993, Jaguar dropped the V12 coupé and Convertible in the States, replacing them with the four-litre models, priced at $49,750 and $56,750 respectively. This was the first time a six-cylinder S had been offered in America, but it was thought the $10,000 reduction in price would tempt people back into Jaguar showrooms. The plan worked and no less than 12,267 six-cylinder cars were sold in the US over the following four years - over seven times the number of V12s sold there in the same period.

Producing 219bhp at 4750rpm in US trim, the AJ6 unit could be mated

The four-litre Convertible for the US 1993 model year. This was the first time a six-cylinder XJS had been offered in the States.

The facelifted XJR-S. For 1993, standard XJR-S colours included Brooklands Green, Flamenco, Morocco Red, Silver Frost and Solent Blue in the UK.

to the ZF four-speed automatic or, at extra cost, Getrag's five-speed manual gearbox. With a compression ratio of 9.5:1, maximum torque was listed at 273lbft at 3650rpm.

On the manual car, with a 3.54:1 final drive and gearbox ratios of 3.55, 2.04, 1.40, 1.00 and 0.70:1, maximum speeds through the gears were 34, 58, 84, 118 and 138mph. 0-60 came up in just 7.4 seconds, while the standing-quarter was covered in 15.5 seconds at a speed of 91mph. The new four-litre model was therefore actually quicker than the old V12. However, for those who still wanted this legendary power unit, it was available in the XJR-S.

The XJR-S
The XJR-S continued alongside the facelifted range during 1991. To coincide with the launch of the new models, the XJR-S received the same tail-end treatment and new rear quarter-lights (this time, the styling work shifted away from TWR and was carried out in-house by Geoff Lawson's department at the Whitley Technical Centre), as well as a boost in power to 333bhp at 5250rpm; maximum torque was listed at 365lbft at 3650rpm.

The XJR-S continued to be built in relatively small numbers until mid-1993 for the home market, by which time it cost a hefty £50,400 - only £500 less than the V12 Convertible, and a massive £17,000 more than the six-cylinder coupé. Weighing in at 1825kg, Jaguar quoted a top speed of 158mph, and a 0-60 time of 6.5 seconds.

In America, the standard V12 XJS was dropped from the range for the 1993 model year, but the six-litre power unit was still available in the XJR-S. This was the first time that the JaguarSport model had been listed in the States, and it made the perfect stablemate for the newly-introduced four-litre machines. However, there would only be 100 models exported to the US - 50 coupés and, for the first *and* last time anywhere, 50 in soft-top form.

Regarding the six-litre XJR-S coupé, *Sports Car International* proclaimed: "The good news is that with the uprated suspension and yeoman service from the engine, the XJR-S becomes an acceptable personal sport/luxury car. Tightening up the suspension and steering have transformed this car. The suspension cure has taken care of the biggest weak spots of the

156

A thoroughbred performance car that traces its lineage directly to Jaguar's highly successful competition efforts, the XJR-S has been developed by JaguarSport, the same group responsible for three FIA World Sportscar Championships since 1987 and resounding victories at the 24 Hours of Le Mans and Daytona 24-Hour in both 1988 and 1990.

Like the Le Mans-winning XJR-12 race cars, this ultimate grand touring machine utilizes a special, custom-made version of Jaguar's incredibly durable overhead cam V-12 engine. Enlarged to 6.0 liters and equipped with a talented duo electronic engine management system, this unit delivers 318 horsepower, ranking it among the world's most powerful road cars available today.

Also in common with the racing Jaguars, the XJR-S sports and complement of aerodynamic styling enhancements. At the front and rear spoiler, body work modifications are

designed to enhance stability at high speeds, while side skirt extensions help smooth the airflow when the car is travelling at speed.

XJR-S		
PERFORMANCE SPECIFICATIONS		
HORSEPOWER:	318 @ 5,200 rpm	
TORQUE:	339 lb-ft @ 3,750 rpm	
TRANSMISSION:	Three-speed automatic	
REAR AXLE RATIO:	2.88:1	
ROAD WHEELS:	Alloy wheels with 225/50ZR16 front tires and 245/55ZR16 rear tires	
WEIGHT:	4,053 lb (coupe) 4,253 lb (convertible)	

X J R - S
WORTHY HEIR TO
THE GLORY OF LE MANS

Of all the invigorating, trend-setting Jaguars that have been built over the last half century, no production model Jaguar convertible has ever been more powerful than the 1993 XJR-S. Jaguar's legendary overhead cam V-12 has been enlarged to 6.0 liters and extensively modified to deliver a lusty 318 horsepower, yet sacrifices none of its

much-heralded smoothness and response.

The suspension, too, has been bred for enhanced performance, providing immediate and accurate response to driver inputs yet maintaining an excellent ride quality over rough road surfaces. Firmer calibration of the springs and shock absorbers, along with massive Z-rated 16-inch tires

and special alloy wheels, further assure that the XJR-S will meet the expectations of the very limited number of

demanding drivers who will experience this unique convertible.

P O W E R   A N D   P R E S T I G E

T H E   X J R - S   C O N V E R T I B L E

The only way Americans could buy a V12 XJS in 1993 was by asking for the XJR-S. In the US it was available in coupé ...

... or soft-top guise. Only 50 of each were built, and were finished in either Signal Red or Jet Black.

XJS: its floaty ride and disembodied steering. Some traditional luxury car buyers may find this suspension a little too jiggly for their tastes, and they are advised to consider the regular XJS. But for anyone who likes to drive a powerful car quickly, this is exactly what the XJS had been lacking for more than 15 years."

The three-speed automatic transmission was the only option (with a 2.88:1 final drive), but with 318bhp and 339lbft of torque on tap thanks to the new Zytek engine management system, the 0-60 time was cut to just 7.1 seconds, while the standing-quarter was covered in only 15.3 seconds before going on to a top speed of 153mph.

Automobile magazine managed to secure an XJR-S Convertible for test, and noted: "The surprise is the chassis, which is astonishingly tight and rattle-free despite the loads imposed by the wide tires. All 1993 XJS models incorporate new stainless steel crossbeams beneath the body to increase torsional rigidity by 40%."

"You can see the limited edition XJR-S is special," said *Motor Trend*. "It

has a unique front and rear bumper treatment with corresponding rocker panels, satin-black grille, headlight bezels, rear spoiler, and the XJR-S logo etched into the rear quarter glass. Lower than the standard XJS, the revised coupé looks far more ferocious, with wide, 8J x 16 inch alloy wheels carrying Z-rated 225/50 front and 245/55 rear Goodyear Eagle tyres. The suspension has been dropped, spring rates raised, and gas-filled Bilsteins installed on all four corners.

"The interior didn't escape JaguarSport's touch, either: Autolux leather trim, plated door jams, the elegant veneer, and a sterling silver plaque [by Asprey] with each car's production number highlight the cockpit. What's missing is sufficient side support in the seats and a left-side dead pedal to help hold you in place with the added grip the suspension and tyres give."

Despite the aging concept, *Car & Driver* declared: "We can't bring ourselves to discard this immense coupé, which, even after 18 years, causes onlookers to sigh with the same envy

they reserve for $107,000 Benz 500SLs. No-one here knows precisely why. But it may explain why Ford paid roughly $2.5 billion just to possess the Jaguar logo."

The only problem was the price, for while the standard 1993 XJS coupé and drophead cost $49,750 and $56,750 respectively, the XJR-S was a massive $73,000 in coupé form, or $80,100 for the Convertible. Nonetheless, most of the production run was presold the moment it was announced.

The XJR-S was dropped when the new XJS was launched in mid-1993, but as a flagship sporting model, it had served the company well. XJR-S production figures are listed below:

Year	Production
1988	205
1989	227
1990	328
1991	65
1992	282
1993	39
Total	1130

158

Left: Although the XJR-S was dead, the S still attracted the specialist attention of tuning and conversion firms. This is the seven-litre Lister Le Mans of mid-1992. With a reduction in new car tax in the UK, the price fell to a still staggering £147,500.

Rear view of the Lister Le Mans.

From these production figures, whilst most of the XJR-S models came with six-litre V12s, it should be noted that 366 of the early coupés were fitted with 5.3 litre engines. Also, of the 1130 cars built in total, 50 were XJR-S Convertibles (34 were produced in 1992, with the remainder being completed in 1993).

A major face-lift

By 1993, federal regulations would have restricted V12 performance to an unacceptably low level for such a big engine. To counteract this, the capacity of the V12 unit was increased to six litres and combined with a low-loss catalyst exhaust system, plus four-speed automatic transmission.

This new drivetrain was fitted to both the XJS and, earlier that year, the new XJ40-type XJ12 and Daimler Double-Six saloons. However, when the six-litre XJS appeared in May, it came complete with a major styling facelift that was applied to both the twelve- and six-cylinder cars. The press release, dated 10 May 1993, read as follows:

Jaguar is announcing a revised XJS range which incorporates significant changes in specification, designed to strengthen the appeal and competitiveness of Jaguar's luxury sports car models worldwide.

The main changes are the installation of the highly acclaimed new six-litre V12 engine, which provides a startling improvement in performance, the availability of a four-speed automatic transmission on V12 models, the in-

The last Series 3 V12 saloons were built at the end of 1992 to make way for the new XJ40 Jaguar XJ12 and Daimler Double-Six, powered by a six-litre version of the legendary V12. Announced at the 1993 Amsterdam Show, they also featured a new four-speed automatic gearbox - something the V12 (and earlier six-cylinder Jaguars) had needed for years.

troduction of a 2+2 XJS Convertible and the fitment of new moulded bumpers and cast alloy wheels, which give both Coupé and Convertible models a more contemporary appearance.

In addition, a new improved sports suspension has been developed as standard equipment for the four-litre and six-litre V12 Coupé models. This new suspension is also to be offered as an option on Convertible models for

the first time. A sophisticated factory fit alarm system is now fitted to all XJS models as standard equipment.

Nick Scheele, Jaguar's Chairman and Chief Executive, underlines the scale of the changes: "The new cars look much fresher, they are nicer to drive and the new six-litre V12 installation gives a tremendous boost to the range. The V12 models now offer really outstanding performance, as well as

The new six-litre V12 engine, seen here installed in the latest incarnation of the XJS. At last, it was mated to a four-speed automatic transmission.

excellent refinement at a value for money price. The availability of the new 2+2 Convertible will broaden the appeal of the range, too. Overall these changes will enable us to compete even more vigorously in the world's luxury sports car markets."

The new V12 engine and four-speed automatic transmission

Jaguar's new six-litre V12 engine was successfully launched earlier this year in the new XJ12 and Daimler Double-Six saloons.

The new V12 engine is derived from Jaguar's long-serving, all-aluminium alloy 5.3 litre V12 which was introduced in 1971. The engine has undergone detail modifications in order to maintain its reputation for providing excellent power, torque and refinement.

The engine displacement for the latest version has been increased to 5994cc from the 5343cc of the previous model. This has been achieved by increasing the stroke from 70mm to 78.5mm. The bore is unchanged at 90mm.

The other main changes include a revised cylinder head, new 'flat-top' piston design, reduced compression ratio, new cylinder liners, revised inlet valves, new camshaft profile for added valvetrain refinement, and a new forged steel crankshaft which replaces the old cast-iron version.

In addition, the new V12 models are equipped with a new low-loss catalyst system, a new torque converter for improved take-off, new twin in-tank fuel pump system, new starter motor, alternator and a new Lucas-Marelli engine management system.

Underbonnet appearance has also been addressed with a revised layout providing a much improved cosmetic appearance.

The new V12 engine provides significant improvements in power, torque and performance. Maximum power is up by 10% to 308bhp (227Kw) at 5350rpm and peak torque is up by 16% to 355lbft (481Nm) at 2850rpm compared to the previous 5.3 litre V12 engine.

As a result of these improvements, the new XJS V12 Coupé and Convertible models deliver even more impressive performance. The XJS V12 Coupé accelerates to 60mph in 6.6 seconds compared to 7.8 seconds for the previous model. Maximum speed is 161mph (260kph) where legislation and road conditions permit, compared to 147mph (237kph) for its predecessor.

160

The new XJS, featuring heavily facelifted styling executed by the late, great Geoff Lawson. I was going to ask Geoff to write the foreword to this book, but his untimely death meant I could not. He will be sorely missed.

Fascia of the four-litre model, complete with five-speed manual transmission.

The biggest news on the Convertible was the 2+2 seating arrangement. If the optional leather trim was specified on the six-cylinder model, it was hard to differentiate between the four and six-litre cars. However, this is a six-litre version, as the switch by the automatic gearbox doesn't have a first gear inhibit marking. The only other clue is the inlay in the burr walnut, which was unique to the V12.

The new V12 Convertible accelerates to 60mph in 6.8 seconds, and has a maximum speed of 160mph (258kph). There are also gains in mid-range performance compared to the outgoing model.

V12 models also benefit from the new GM4L80-E four-speed electronic automatic transmission which uses the same geartrain as its predecessor, the GM400, for first, second and third gears, but an extra set is added to provide the fourth overdrive ratio. The new torque converter contains a lock-up clutch for added economy. A limited-slip differential is standard on all XJS models.

The new gearbox has 'Sport' and 'Normal' modes of operation with a first gear inhibit mode incorporated in

the 'Normal' mode. When the driver selects 'Normal' mode, the car will automatically start in second gear, unless the throttle is opened wide when first is engaged.

In 'Sport' mode the car always starts in first gear. 'Sport' mode is more sensitive, readily enabling part throttle kickdowns so that the more enthusiastic driver can experience and fully enjoy the much improved performance of the new V12 models.

The new four-speed electronic transmission is much more refined than its GM400 three-speed predecessor, too. The GM4L80-E has a Transmission Control Module (TCM) with features that give much improved shift quality. These include the ability to communicate electronically with the

Lucas-Marelli engine management system to trigger a reduction in engine torque during shifts, which enhances refinement. The TCM also has a self-diagnostic capability which can record evidence of malfunction. This data can be retrieved from the TCM using the Jaguar Diagnostic System at Jaguar franchised dealers during regular servicing.

The ZF automatic transmission on four-litre models also incorporates a first gear inhibit.

Exterior changes
Exteriors of both the XJS Coupé and Convertible have a more contemporary look, with new, colour-keyed moulded bumper covers [with chrome-plated upper blades], new rectangular style

The new Convertible with hood up. Like all Jaguar soft-top models since their introduction in 1988, the rear screen was glass and featured a heated element.

exhaust tailpipe finishers and new-style road wheels.

An attractive new 7 x 16 inch, five-spoke, cast alloy wheel is fitted to all versions, replacing the 6.5 x 15 inch alloy wheel fitted as standard to previous models.

The new 2+2 convertible model

The revised XJS range also includes a 2+2 Convertible model which becomes the standard condition for the Convertible model in the UK, Europe and most overseas markets.

The 2+2 Convertible has a unique body-in-white created by welding a sub-assembly on top of the existing underframe. This forms the new rear seat pan and seat back in place of the rear stowage box area on the two-seater models.

The new rear compartment area features two occasional seats, two seat belts, new carpeting and new trim.

The rear windscreen is slightly shallower than on the two-seater version, which enables the hood and screen to be folded behind the new rear seat whilst maintaining the stowed hood stack height of the two-seater. The hood stowage cover material changes from cloth to Ambla. The Ambla is colour-keyed to interior trim to give a more co-ordinated appearance. To accommodate the rear seats, the hydraulic hood pump has been relocated to the boot.

Chassis modifications

All four- and six-litre Coupé models are equipped with a newly-developed sports suspension as standard, which provides an even better ride and handling compromise. A softer 'Touring' suspension is available at no extra cost.

Changes to the sports suspension include Bilstein dampers, revisions to the spring rates and front anti-roll bar specification, and deletion of the rear anti-roll bar. In addition, cars with sports suspension have low profile Pirelli P600 225/55 ZR16 sport tyres fitted to the new 16 inch alloy wheels.

Four-litre and six-litre V12 Convertible models are equipped with the softer 'Touring' suspension as standard, although the sports suspension option is now available to Convertible buyers for the first time. On 'Touring' suspension models a new 'Comfort' tyre, the Pirelli P4000E 225/60 ZR16, is specified.

An alloy spacesaver spare wheel will also be available as a no-cost option. The wheel size is 3.5 x 18 inches [and comes fitted] with a 115/85 R18 tyre. The existing 16 inch forged lattice alloy wheel continues as a cost option in the new range.

Other changes include re-siting of the rear brakes outboard and fitment of new calipers for improved braking performance. All XJS models will benefit from the new ZF steering rack introduced earlier this year, which provides better "on-centre" feel and progression.

Interior changes

There are a number of changes to the interior, designed to improve occupant comfort and convenience.

These include new sunvisors which incorporate illuminated vanity mirrors and new interior lights. The effectiveness of the air conditioning system has been improved with revisions to the air vents, ambient air sensor and motorised aspirator. These changes provide improved temperature stability.

All models feature new warning chimes which will sound if the driver's seat is occupied but the seat belt is not being worn. The chimes will stop after six seconds.

New vehicle security system as standard equipment

A fully-integrated, factory-fitted alarm system has been developed which will be fitted to all models in the XJS range as standard equipment. All models in the Jaguar product range now have alarm systems fitted as standard.

The new XJS alarm system is remotely operated by RF using a hand-held transmitter which also operates the central locking system. The transmitter will also activate a headlamp illumination feature. The security system senses the perimeter of the car, i.e.

Specifications

Engines

	6.0 V12	4.0 AJ6
Bore	90mm	91mm
Stroke	78.5mm	102mm
Capacity	5994cc	3980cc
Compression ratio	11.0:1	9.5:1
Combustion chambers	May HE	Pent-roof
Maximum power (bhp @ rpm)	308 @ 5350	223 @ 4750
Maximum power (Kw @ rpm)	227 @ 5350	166 @ 4750
Maximum torque (lbft @ rpm)	355 @ 2850	278 @ 3650
Maximum torque (Nm @ rpm)	481 @ 2850	377 @ 3650
Ignition type	Marelli elec.	Lucas elec.
Fuel-injection type	Lucas elec.	Lucas elec.

Transmissions

Manual gearbox (for 4.0)	Getrag 290 five-speed
Automatic gearbox (for 4.0)	ZF 4HP-24
Automatic gearbox (for 6.0)	General Motors GM4L80-E

Ratios		Getrag	ZF	GM
	1st	3.55	2.48	2.48
	2nd	2.04	1.48	1.48
	3rd	1.40	1.00	1.10
	4th	1.00	0.73	1.00
	5th	0.70	-	-
Axle ratios		3.54:1	3.54:1	2.88:1

Braking

Electronically-controlled, anti-lock braking system with yaw control. Hydraulic power-assisted four-wheel disc brakes, ventilated at front. Safety split front and rear hydraulic circuits incorporating fluid loss sensor warning. Hand operated mechanical parking brake on rear wheels.

Steering

Rack-and-pinion power-assisted steering. Energy absorbing steering column. Four-spoke (SRS) airbag steering wheel with four-position vertical adjustment.

Suspension

Front: Fully-independent with twin wishbones, coil springs and telescopic dampers. Anti-roll bar. Anti-dive geometry providing longitudinal stability under heavy braking. Sports damper settings on Coupés.

Rear: Fully-independent with lower transverse wishbones and driveshafts acting as upper links. Radius arms, twin coil springs and telescopic dampers. Sports damper settings on Coupés.

---

the opening of the doors and boot. Visual deterrents include a flashing red indicator on the centre console when armed, and warning labels on the side windows.

The system activates both visual and audible alarm signals, involving side and taillights and a dedicated alarm siren.

Comprehensive standard specification

The new XJS range maintains Jaguar's reputation for offering outstanding value for money. In addition to its electrifying performance, the new XJS six-litre V12 Coupé also offers as standard: a driver's side airbag, anti-lock brakes, catalyst exhaust system, a limited-slip differential, cruise control, four-speed automatic transmission, sports suspension, powered leather sports seats with memory, burr walnut veneer with inlays on the doors and rear quarter panels, leather-trimmed centre console, air conditioning and a custom-designed audio system [a CD Autochanger was listed as an option] amongst many other features, all at a price which is considerably below those of its principal competitors.

Press reaction

UK prices for the new XJS range were as follows: the XJS 4.0 was listed at £33,600, the 4.0 Convertible was £41,400, the V12 Coupé cost £45,100, and the flagship model, the XJS V12

External dimensions

	Coupé	Convertible
Overall length	4.82m (189.8in)	4.82m (189.8in)
Overall height	1.24m (48.9in)	1.28m (50.2in)
Overall width	1.79m (70.6in)	1.79m (70.6in)
Wheelbase	2.59m (102.0in)	2.59m (102.0in)
Front track	1.49m (58.6in)	1.49m (58.6in)
Rear track	1.50m (59.2in)	1.50m (59.2in)
Ground clearance	120mm (4.7in)	120mm (4.7in)

Weights

	4.0 man.	4.0 auto.	6.0 auto.
Coupé (lb/kg)	3760/1705	3760/1705	4101/1860
Convertible (lb/kg)	4035/1830	4035/1830	4377/1985

The new XJ-S V12 models - comparative data:

	XJS 6.0 V12	XJS 5.3 V12
Power (bhp DIN)	308 at 5350rpm (227Kw)	280 at 5550rpm (209Kw)
Torque (lbft)	355 at 2850rpm (481Nm)	306 at 2800rpm (415Nm)

Performance

XJ-S V12 Coupé:

0-60 mph (secs)	6.6	7.8
Maximum speed (mph)	161 (260kph)	147 (237kph)

XJ-S V12 Convertible:

0-60 mph (secs)	6.8	8.1
Maximum speed (mph)	160 (258kph)	143 (230kph)

Fuel economy

XJ-S V12 Coupé:

Urban - mpg (litres/100km)	13.2 (21.4)	14.2 (19.9)
56mph (90kph) - mpg (litres/100km)	26.4 (10.7)	25.7 (11.0)
75mph (120kph) - mpg (litres/100km)	22.6 (12.5)	21.1 (13.4)
Combined - mpg (litres/100km)	18.6 (15.2)	19.1 (14.8)

XJ-S V12 Convertible:

Urban - mpg (litres/100km)	13.2 (21.4)	13.7 (20.6)
56mph (90kph) - mpg (litres/100km)	26.4 (10.7)	26.6 (10.6)
75mph (120kph) - mpg (litres/100km)	22.6 (12.5)	21.9 (12.9)
Combined - mpg (litres/100km)	18.6 (15.2)	19.1 (14.8)

Convertible, was £52,900 - a far cry from the prices of 1975. At the time of the new car's launch, the Jaguar range started at £26,200 for the XJ6 3.2 and went up to £51,700 for the Daimler Double-Six.

The excellent but sadly short-lived *Complete Car* secured a V12 coupé for long-term appraisal. The magazine was not so happy with some of the switchgear, the cramped rear seats, "poor ride" and general expense of running the vehicle, but was very compli-

mentary about the engine. It was both comfortable and relaxing, and the magazine went on to describe the S as a "fine means of transport."

As for the handling, it was stated: "Soggy handling and a very poor ability when cornering used to accompany the silky XJ-S image. But as the model evolved, Jaguar engineers revised the car and now it corners well but is much more intimate with the road surface. I'm not sure this is a good thing, and there is still a vague feeling when steer-

ing straight-ahead.

"The XJS is no sports car, after all. Handling is not really something a gentleman has any interest in, and the XJS is, above all, a car for gentlemen."

Writing for *Autocar & Motor*, Gavin Conway said: "The big news behind Jaguar's first-ever four-seater XJS convertible lies under the bonnet, where a more potent six-litre V12 now resides. This engine, which first appeared in the XJ12 saloon, also graces the revised XJS coupé. It produces 308bhp,

The Convertible and coupé together. Note the lack of a 'V12' badge on the soft-top's bootlid, one of the few external clues that show this is a four-litre model.

compared to the old 5.3 litre V12's 280bhp. Torque is up, too, from 306lbft to a truly breathtaking 355lbft. The bad news is that those power figures are big enough to step on the toes of JaguarSport's brawny XJR-S. Jaguar has therefore decided to drop the XJR-S from its line-up.

"The XJS V12 puts its leviathan power to the rear wheels via the same General Motors four-speed automatic gearbox that made its Jaguar debut in the V12 saloon. The four-speeder has a switchable mode ... But the real trick to this transmission is the way that engine torque is electronically reduced during gearchanges to provide unbelievably fluid shifts.

"When Jaguar started building the first truly convertible XJS five years ago, it was a two-seat affair. But then the US, one of Jaguar's most important markets, began to complain. They were initially happy with two seats,

then they changed their minds,' says Jaguar about the fickleness of US customer loyalty to the open-top XJS. With 65% of all XJSs sold in America being convertible, it was a complaint that Jaguar could not ignore. In the UK, that ratio is reversed, with the coupé hugely outselling the soft-top."

Conway concluded: "While £52,900 may sound steep for so ancient a piece, consider that the only other V12 open-top cruiser on sale, the Mercedes-Benz 600SL, costs an additional £44,000." One of the car's first customers was the Duchess of York, who took delivery of her gold V12 Convertible in July.

Other events
The Aston Martin DB7 made its debut in March 1993, but it wasn't until some time after this date that it hit the showrooms. Based around the XJS and powered by a supercharged 3.2

litre Jaguar AJ6 engine (the supercharged version of the Jaguar XJ6 saloon - the XJR - employed the four-litre unit), the coupé's styling bore a striking resemblance to the doomed F-type.

The DB7 was built at the JaguarSport factory in Bloxham (where XJ220 had been produced), with the first customer cars, priced at almost £80,000, leaving the works in summer 1994. With 335bhp on tap, performance was all that the F-type should have had. Still in production as we entered the new millennium, it was later announced that some DB7 models would be endowed with V12 power.

In the meantime, in America, with the XJR-S gone, the standard V12 coupé and soft-top returned to the US line-up for 1994. There were now four XJSs sold in the States, all the updated models launched originally in May 1993, and press reaction was "almost

The AJ6 engine for 1994. Note the new oil filler location - an identifying feature for the year.

Part of the all-new X300 range, with a supercharged XJR in the foreground, an XJ12 on the right, and a Daimler Six in the background.

uniformly positive," as a member of the Jaguar Cars Inc. staff put it. Available from August 1993, 1994 model year XJS prices started at $51,950 for the four-litre coupé, rising to $79,950 for the V12 Convertible.

On the home market, in September 1993 it was announced that 1994 MY cars would have a new-style air conditioning control panel to give "im-

proved operation" and additional features (the system also became CFC-free at the same time), a new screen-mounted rear-view mirror, and twin airbags were fitted as standard. The AJ6 engine was given a new camshaft cover and inlet manifold, and a gearshift interlock was incorporated into the automatic transmission for added safety. For the soft-top models, a bur-

gundy-coloured hood was made available, increasing the choice of shades to five. In the UK, prices for the 1994 XJSs ranged from £35,400 to £55,300.

Initial figures released for the first quarter of 1994 were very encouraging, with registrations of the four-litre Convertible up by almost 80% in the States, thanks to strong sales in Florida and California.

On a sadder note, it was announced that Venture Pressings was to close. This had been a joint project between Jaguar and GKN, the new Telford plant taking the place of the Rover press shop in Swindon. It opened in November 1990, mainly to supply XJS body pressings (all of the 1991-93 XJS panels came from Venture Pressings, although assembly continued at the refurbished Castle Bromwich site), but Ford simply didn't need the factory, as it had its own facilities.

In autumn 1994, the all-new X300 XJ Series, including four-litre sixes and six-litre V12 models in both Jaguar and Daimler guise, made its debut at the Paris Motor Show. The X300 models featured more sculptered styling than the outgoing XJ40s, and received critical acclaim from all over the globe.

Having loaned a Double-Six from the factory, I wrote in my history of the Daimler and Lanchester marques: "The author can confirm that the X300 is without doubt one of the finest cars in the world, and will go down in Jaguar history as a landmark for the company. With effortless performance, exceptional refinement and more than reasonable fuel consumption for a six-

In summer 1994, the AJ6 was replaced by the more refined AJ16 (Courtesy Paul Skilleter).

litre V12 engine, it is probably the best car ever to wear the exclusive Daimler grille, and is much better than the XJ40 and anything Crewe or Stuttgart currently has to offer."

At the time of the X300 launch, the new saloons ranged in price from £28,950 for the XJ6 3.2 to £59,950 for the ultra-luxurious Daimler Double-Six. By this time, the price of the XJS had risen considerably. The four-litre coupé was £36,800, while the Convertible with the same engine was listed at £45,100. As for the V12 models, the coupé was priced at £50,500, and the soft-top came in at a massive £58,800.

The 1995 model year

Having been announced in early June 1994, the biggest change for the 1995 model year XJS range concerned the four-litre models, which were given the new AJ16 unit. As *Performance Car* noted enthusiastically: "There's a new

cylinder head and block, revised cam profiles, a higher compression ratio with new pistons, lightweight valvegear and new sequential fuel-injection, all designed to give the engine greater performance, durability and refinement. A new throttle system, induction manifold and exhaust helps to liberate an extra 18bhp over the old engine: it now pushes out 241bhp at just 4700rpm, and 282lbft of torque at 4000 rpm.

"Nor are the changes restricted solely to the mechanicals, as the engine now looks different under the bonnet. Pleasingly, there's a handsome new silver cam cover scripted in green with the word Jaguar in the corporate typeface that takes your mind back to the company's competition past. Very Jaguar and sure to be a hit with our American friends.

"Other changes include new seats with integral headrests and updated

trim materials, and there are twin airbags and a new, more comprehensive sound system. Smart, shiny finish five-spoke alloys and colour-keyed door mirrors and headlamp surrounds pick out the exterior differences, and for the first time customers can specify chrome-plating for the alloys as an extra. And isn't it good to see that at last the old car spares shop chrome exhaust trims have given way to more aesthetically pleasing oval stainless ones?"

The V12 models received new 20-spoke alloy wheels (the chrome-plated five-spoke items could be specified as an option), a headlamp power wash system, a black grille instead of the body-coloured item found on the smaller-engined S, an enhanced interior and, for America, a rear spoiler which incorporated a high-mounted brake light; chrome mirrors and headlight surrounds were retained on the

An AJ16-powered Convertible, distinguished by the body-coloured door mirrors and headlight surrounds. The contemporary four-litre coupés received the same treatment, whereas the V12s retained chrome mirrors and headlight trim.

The six-litre XJS coupé from June 1994. The new alloy wheels really set off the car, but other detail changes of note include 'V12' badges on the front wings and the gold-coloured growler.

XJS COUPE EXPRESS TRAVEL, SPORTING STYLE

High performance and superlative luxury mark out the Jaguar XJS Coupé as one of the most coveted of sporting cars.

For 1995, the 4.0 litre coupés offer a remarkable advance in performance, refinement and economy, thanks to their new AJ16 engine. The optional four-speed automatic transmission also gains a low-inertia torque converter for effortless performance off-the-line. Top speed rises to 147 mph (where permissible), with acceleration from 0-60 mph in 6.9 seconds with the standard five-speed manual gearbox. Comparable figures for the uncannily

smooth XJS V12 Coupé are 0-60 in 6.6 seconds, with a maximum speed of more than 160 mph. Equally impressive is fuel economy of 37.9 mpg at a constant 56 mph for the 4.0 litre XJS Coupé automatic transmission: a 5.9% improvement over previous models.

Exterior detailing reflects the individual style of the two XJS models: the 4.0 litre coupé gains colour-keyed door mirrors, grille and headlamp surrounds while the V12 provides a restrained glint of chromium for headlamp bezels to contrast with the black grille.

Inside the luxuriously-appointed 2+2 cabin are re-styled seats and even higher quality trim, making extensive use of soft autolux leather in the V12 models.

	XJS 4.0 Coupé	XJS V12 Coupé
Acceleration - 0-60 mph (0-100 km/h) (62 mm./min.)	6.9 (7.6)/7.6 (8.1)	6.6 (6.9)
Top speed - mph (km/h)	147 (237)	161 (260)
Max power DIN BHP (kW) @ rev/min	237 (177) @ 4700	304 (227) @ 5250
Max torque DIN lb.ft (Nm) @ rev/min	282 (382) @ 4000	355 (481) @ 3850

XJS coupés for the 1995 model year. The interior shot at the top shows the seating's ruche leather centre panels, introduced on the V12 in mid-1994 (the optional leather trim on the four-litre cars had smooth centre panels).

six-litre cars. But, as *Road & Track* pointed out: "What really made the experience of driving the car different was the sense of immediacy. There was no lag anywhere in the whole operation or control of the automobile. Steering, braking, acceleration and tracking were so instantaneous as to seem prescient.

"At the engineering level, this kinetic response occurs because the horsepower rating of the 1995 XJS six-litre V12 has been improved to 301bhp at 5400rpm (nearly a 10% increase over 1994) and the torque output now measures a complementing 351lbft at 2800rpm."

Jaguar's Roger Putman said: "The XJS V12 buyer demands the ultimate in performance, refinement, technical sophistication and luxury. With the

latest refinements, the XJS V12 remains the definitive luxury Grand Tourer."

End of an era

During the first half of 1995, sales of the four-litre soft-top had been extremely strong in the US; in fact, some 34% up on 1994 figures. However, from May 1995, the V12 XJS models were available to special order only and, at the end of the year, further tightening of US regulations led to the complete withdrawal from the American market of the V12-powered XJS (only 413 V12s were sold Stateside in 1995).

The final run of four-litre models were given the 'Celebration' title to commemorate 60 years of the Jaguar marque, and were distinguished by

their new, almost flush, 11-spoke alloy wheels (similar to those of the last XJR-S, except the latter used an 8J x 16 wheel instead of the 7J x 16s found on the latest XJS models). They were listed for the 1996 model year in both Europe and the United States.

Announced on 10 May 1995, the press release read: "The new Celebration Coupé and 2+2 Convertible feature the highly-acclaimed twin-camshaft, four-litre AJ16 engine from the new XJ Series saloon, four-speed electronic automatic gearbox, cruise control and distinctive 'Aerosport' diamond-turned alloy wheels. The luxurious interior has unique sapwood walnut veneers, leather trim, a wood-rimmed steering wheel and a matching wooden gearknob. A new braking sys-

A UK specification Celebration Convertible, pictured at the time of announcement. No less than 17 shades of paintwork were available, with five choices for hood colour. V12 cars were built to special order only for 1996, although very few customers took up the option.

Interior of the Celebration models: note the new steering wheel.

172

A home market Celebration coupé from the 1996 model year. For the Celebration models - which were only ever sold in four-litre guise - distinctive 'Aerosport' wheels were fitted, and it was decided to give the headlight surrounds and door mirrors a chrome finish.

tem, first introduced on the new XJ Series saloon, is now fitted to the XJS, yielding improvements in pedal effort, pedal sensitivity and feel.

"With the four-litre Coupé priced at £38,950 and the four-litre Convertible at £45,950, the Celebration XJS models reflect more than ever that unrivalled blend of Jaguar style, prestige, performance, luxury and affordability." The release went on to describe the "booming demand" for the XJS, particularly in America. But everyone could see that time was running out for the S. In April 1996, production of the XJS finally ceased in order to make way for the XK8 (see chapter seven). The last cars - a red four-litre soft-top and a light blue V12 coupé - rolled off the line on 4 April to take their places in the JDHT collection. Sales of the final XJSs went extremely well,

with the last cars sold in the UK in May, whilst American dealers held on to the remaining few as they were commanding a premium - 27 V12s and 2,115 six-cylinder models were sold in the States in 1996.

It was so sad to see the end of this great Grand Tourer, especially as Jaguar had improved the car so much over the years. Gavin Conway, then with *Autocar*, summed up XJS motoring with the following eulogy: "It has a wonderful ride, a gloriously revived engine and a cabin that makes even minor trips an occasion. It will make you feel better about yourself, I promise."

But the XJS was not the only part of Jaguar's heritage to fall by the wayside. With the latest six-cylinder engine delivering similar levels of refinement, but with far superior fuel

economy, ever since the AJ16 had been launched in 1994, the days of the V12 were numbered. In a press release dated April 1997, the inevitable was stated: "After a 26 year reign as Jaguar's flagship engine, production of the Coventry-based company's legendary V12 has ceased. In a total production run of 161,996 engines, Jaguar's trend-setting V12 powered a plethora of saloons and sports cars with outstanding international success on both road and race track. The final V12-engined car, a six-litre Jaguar XJ12 saloon, rolled off the production line on 17 April and takes its rightful place in the company's Heritage Trust collection."

Commenting on the contribution of the V12 engine, Nick Scheele, Jaguar's Chairman and Chief Executive, said: "At its launch in 1971, the V12

The Jaguar XJS Convertible is an exercise in exhilaration. Power is derived from a 4.0-litre, 6-cylinder, 237-hp engine that helps the XJS to devour the straightaways and snake through the curves with sporting dash.

The XJS cockpit boasts a new custom wood and leather steering wheel and walnut gearshift knob as standard features, complementing its already lavish interior amenities.

Meanwhile, you are nestled in lordly comfort in a leather and wood-trimmed cockpit, fitted with amenities worthy of a luxury automobile: a premium, multi-speaker audio stereo system; 6-way adjustable front seats with power lumbar support; and--if top-down cruising isn't in the forecast--automatic climate control.

engine astounded the motoring world with its power, refinement and flexibility. For nearly three decades it proudly fulfilled its role as flagship of the Jaguar engine range.

"Increasingly stringent emissions regulations and changing customer tastes demanded a new breed of engine that delivers both refined power and excellent economy, while satisfying all legislative requirements. The new AJ-V8, launched last year to critical acclaim in the XK8 sports car, has successfully assumed the V12 mantle."

The four-litre AJ-V8, introduced in mid-1996 and making its debut in the XK8 sports car, was to power the entire Jaguar-Daimler range. Not only

had the XJS gone, but the legendary V12 engine had, too. It was truly the end of an era.

The final Celebration model in America. Despite a somewhat rocky start to its career, the XJS went on to become Jaguar's best-selling sports car. Well over 115,000 were sold by the time production came to an end. Just over half of these were exported to the USA.

7

THE FUTURE

"The final irony was that the money spent on building the three concept cars allowed us to judge them and find them wanting, and early in 1990 the programme was cancelled. A harsh decision for all those who'd put so much into it, but an easy one once cold objectivity was applied."
- **Peter Taylor, Jaguar's former Vehicle Proving Engineer, recalling the XJ41 project.**

At first, it was envisaged that the XJ41 and 42 would be a successor to the legendary E-type, and sold alongside the XJ-S. However, as the project developed, it quickly became apparent that the F-type would be more like a direct replacement for the S in marketing terms.

The F-type

Speculation surrounding the so-called F-type was rife throughout the 1980s. Based on the forthcoming XJ40 saloon, it would have signalled a return to open-top motoring for the Coventry company, as both a targa version (XJ41) and a full convertible (XJ42) were planned.

The project started in mid-1980. The XJ41 and 42 had been mentioned in the XJ40 presentation, which stated that the saloon's floorpan and components would be employed to produce a new sports car range "at minimum cost and within a relatively short time period."

Following approval of XJ40 (the fourth generation XJ6) in July of 1980, a number of Keith Helfet's design sketches were eventually presented to Michael Edwardes and the rest of the BL board. The designs were truly beautiful, and the new models were approved in July 1982. By the following January, a glassfibre mock-up was already being assessed in a Los Angeles dealer clinic.

A second glassfibre prototype was duly constructed, featuring a slightly lower roofline and waistline, and some exquisite detailing around the nose and tail. Sir William Lyons, who was still visiting the styling studios on a weekly basis, despite having officially retired, was said to be delighted with it, and again, it was shipped to LA for appraisal.

As XJ41 and 42 took shape, performance figures were established. A four-litre car was proposed for America, but even the 3.6 litre AJ6-powered model for the rest of the world was said to be capable of 159mph and a 0-60 time of 6.6 seconds with a manual gearbox. Aerodynamics certainly helped (the Cd was only 0.30), as did the weight; it was predicted that XJ41 would be up to 15% lighter than the XJ-S.

The launch was tentatively set for March 1986, but with XJ40 running behind schedule, as the months passed this looked increasingly doubtful. A new date of October 1988 was set - two years after XJ40 - and development continued.

As the F-type's contemporaries moved forward, a 330bhp, twin-turbo, four-litre unit was duly specified to keep the new car ahead in the performance stakes, endowing the Jaguar with a 170mph top speed and a 0-60 time of under six seconds. However, in 1985, responsibility for the project was moved to the recently-formed New Vehicle Concepts (NVC) department, and the car underwent some dramatic changes.

The modifications put forward by the NVC team made the F-type more complex, and more and more weight was added to the vehicle. Eventually, it was decided that four-wheel drive would be needed with the twin-turbo car, but a number of problems dictated

Two of the later XJ41 and XJ42 styling bucks in storage at Browns Lane. The author's friend and mentor, Paul Skilleter, can be seen at the back of the convertible.

E-type meets F-type almost a decade after the latter project was cancelled.

A rear three-quarter view of the convertible.

that the body should be made wider. By now, mechanically, XJ41 was starting to move away from its XJ40 origins, thus losing many of the cost benefits outlined in the initial proposal. Furthermore, it was wider and heavier than the XJ-S but with very little in the way of luggage space.

During summer 1988, XJ41 was still being displayed at styling clinics, so there was no chance of an October launch. At least one new Jaguar appeared at the NEC that year - XJ220. The XJ220 supercar was also designed by Keith Helfet, and attracted worldwide attention. With a buoyant financial climate, it later went into limited production.

In the meantime, Karmann, that famous coachbuilding concern, produced three vehicles for Jaguar to test in 1989: a normally-aspirated convertible in red (with automatic transmission, incidentally); a turbocharged dark blue targa and a turbocharged silver targa which was used for the high-speed test runs at Nardo in Italy, where it easily clocked 175mph.

Autocar & Motor published pictures of the so-called F-type in the 27 September 1989 issue. Considering how well the factory managed to keep the project under wraps, the magazine was extremely accurate in many of its statements.

Ultimately, the project was officially cancelled in March 1990. The F-type had dragged on for too long (it was now scheduled for a 1994 model year introduction), and the competition had not only caught up but already surpassed the car in many areas. Had it

The interior of XJ42. Note the automatic transmission.

been launched in early 1986, as originally envisaged, it could have taken the world by storm. Although the styling was still as fresh as it had been a decade earlier, as a package it had too many compromises.

The author was lucky enough to be a passenger in the twin-turbo, targa-top coupé, and it is indeed a very quick car. The overwhelming memory, however, was how cramped the cockpit was, and how little space there was for

XJ41 with the XKR coupé. The styling similarities are very evident in this picture, but time simply ran out for the F-type.

The car that replaced the XJS - the XK8, seen here with a Series 2 E-type; the two cars are separated by a quarter of a century.

Rear end of an early XK8.

luggage - by comparison, it made the boot in my Maserati look like it was designed for parcel delivery!

The XK8

Jaguar's new AJ-V8 engine was announced in June 1996, a few months ahead of the XK8. The four-litre unit produced 290bhp and 290lbft of torque, and was actually built at the Ford plant in Bridgend. Linked to a five-speed automatic, it would soon replace Jaguar's existing six-cylinder and V12 power units.

The first model to receive the new V8 was the XK8 sports car which, having made its debut at the 1996 Geneva Show, went on sale in the UK at the beginning of October, filtering onto export markets a few weeks later.

The XK8 project was born in the opening months of 1990, once it had become clear that XJ41/42 was dead. With two-thirds of the investment needed to develop a brand new car required to bring the XJS up-to-date (to meet forthcoming regulations, and so on), a completely new vehicle was the obvious answer. Serious development began in mid-1992 when the X100 codename was adopted, although it would be another two years before Ford gave it the official stamp of approval.

In complete contrast to the XJ41, the X100 project, overseen by Bob Dover (now the boss at Aston Martin), went exceptionally smoothly, taking just 30 months to go from programme approval to salesroom. The tasteful styling was reminiscent of XJ41, and was the responsibility of Geoff Lawson's

department at Whitley, to where Jaguar moved its development centre in 1988.

As Nick Scheele, Jaguar's Chairman, said at the time of the launch: "The XK8 marks the beginning of a new era in Jaguar sports car history. I am confident that XK8, with its outstanding beauty and all-new powertrain, will re-establish Jaguar as a producer of high quality, exciting and technically innovative sports cars."

Launched at £47,950 for the coupé and £56,850 for the convertible, it is enough to say that first quarter sports car sales were the best in Jaguar's

Right: The new 290bhp AJ-V8 power unit in situ in the XK8 engine bay.

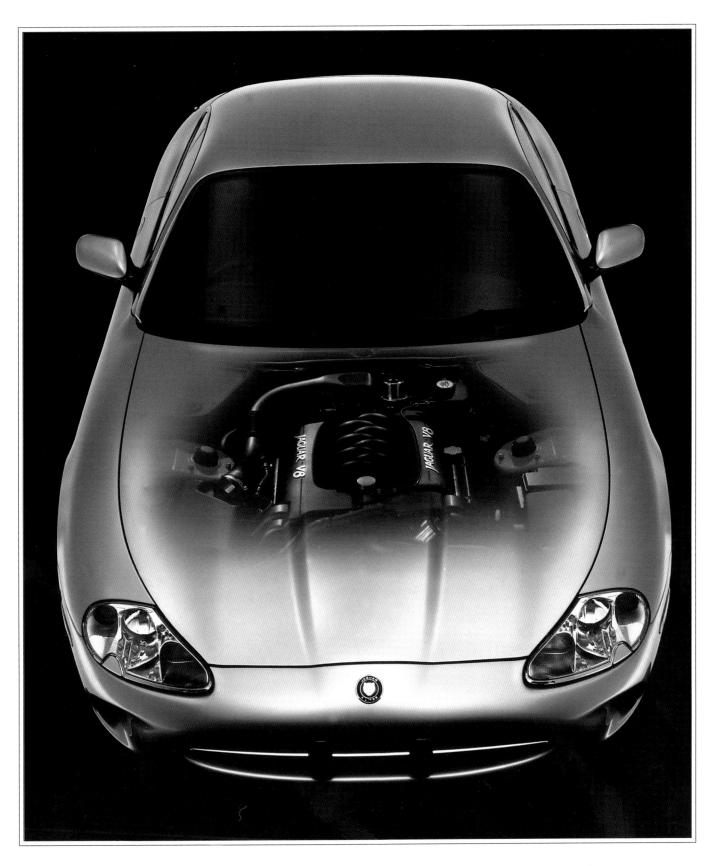

Interior styling of the XK8 was a bold departure from that of the XJS. Somehow, though, it was still typically Jaguar.

Geoff Lawson, Jaguar's Styling Director, with one of the many awards his department has received in recent years. This is an award from Italy for 'the world's most beautiful luxury sports car'.

The convertible version of the XK8, pictured here at the time of its launch.

The mighty XKR with an earlier Jaguar masterpiece - the XK120. Unrestricted, it is thought the supercharged XKR could top 175mph!

A glimpse into the future of sporting Jaguars? The Keith Helfet-designed XK180 was unveiled at the 1998 Paris Salon, and is powered by a 450bhp supercharged V8. Sadly, rumours regarding the possibility of XK180 entering production are being vigorously denied.

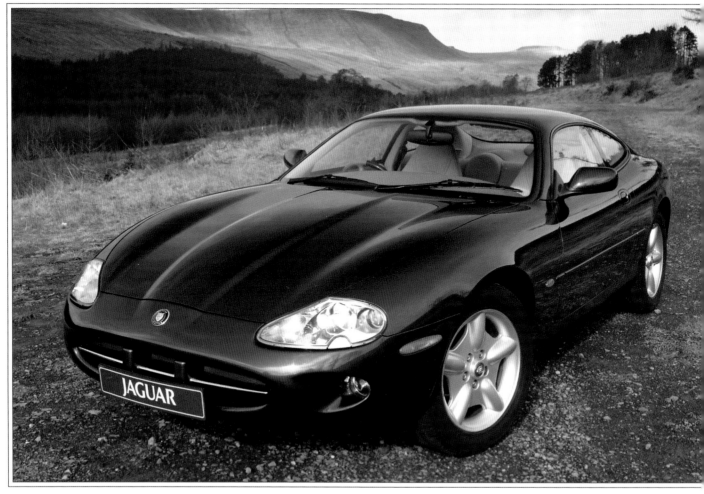

A 1999 model year XK8 - part of a new Jaguar line-up which looks set to assure the Coventry firm's future.

history, and represented a 257% increase compared with the first three months of 1996.

The supercharged XKR models - again available in coupé or convertible form - made their debut at the 1998 Geneva Show. A couple of months later, technical details were disclosed. Power came from the XJR saloon's supercharged AJ-V8 unit, the four-litres producing 370bhp and 387lbft of torque

- enough to give a 0-60 time of 5.2 seconds and a restricted 155mph, despite the automatic transmission.

Prices started at £59,300 for the XKR, with the convertible model commanding another £7000. As *Jaguar World* put it: "There's probably no finer - or faster - way to drive to the South of France, and no similar car with better build quality. Reliability and creature comforts are world class." Sir William

would have been very proud.

The XK8 and XKR are truly magnificent vehicles, as anyone who has driven one will testify. They are as attractive as any of the cars that have sported the Jaguar badge; they are fast, safe and reliable, but have a very different character to the XJ-S. The XK8 is dynamically superior in every department, but the author still has a soft spot for the old S ...

The XJ-S has had a bad reputation for breaking down and rusting for many years. However, in the author's experience, reliability problems nowadays are rarely related directly to the factory. This view was echoed by one of the key people involved with the Nissan Infiniti Q45 project, when he said "I love Jaguar's taste, but their parts manufacturers have let them down."

I remember seeing a bumper sticker in America stating: "Why do the British drink warm beer? Because their refrigerators are made by Lucas." There was always a funny side, but frequent breakdowns and hefty repair bills are rarely amusing. What's more, the reputation the marque acquired during the 1970s for poor reliability was hurting sales. This reputation wasn't unfounded either; in one particular European country, the average amount spent on warranty work was about 12% of the retail price - clearly things were going from bad to worse.

Build quality was another problem, however, and for that the blame had to go to Browns Lane. Maurice Ford, an old friend with many years' experience in the industry, once said, "The Japanese could throw a door from ten yards and it would fit. You could be messing around for ages trying to get an early Jaguar door right, and then you'd have to adjust it again once all the electrics went in." The Japanese spent their time and energy on tooling first, rather than letting the men on the track make the best of a bad job.

By the time XJ-S production came to an end, production methods were completely different to those of 1975, and both build quality and reliability were remarkably good. The following is a brief guide to buying, restoring and running an XJ-S.

Body

As with so many vehicles, the body is the main item of concern with the XJ-S. The 1970s and early 1980s seems to be a period when virtually every car manufacturer in every country (with the notable exception of Germany's more prestigious marques) had a bad rust problem. Everyone talks about the Italian cars of the time rusting but, having been in the trade, the author knows that they were far from alone.

Look for rust in the front and rear wheelarches, the leading edge of the front wings, and where the base of the front wing meets the sill; the sills themselves also deserve careful inspection, and look at the front and rear valances and where the latter joins the rear wings. Check the door bottoms and inside of the bootlid; also at the base of the rear window and the innermost sections of the so-called flying buttresses. Lift the carpets and be sure to inspect the floors and inner sill area, particularly around the rear jacking points where the rear radius arms attach to the body.

Apart from searching for rust (on earlier cars, you probably won't have to search too hard to find it!), look for ripples in the bodywork, poor-fitting and misaligned panels, and differences in the paintwork - all standard practice to establish whether the car has been in an accident. If it has received a

recent respray, has it been 'blown over' for a quick sale, and what is hidden underneath? It is often better to buy a nice original car than one that looks shiny.

As *Autocar* rightfully said: "The XJ-S has an immensely strong monocoque, and no normal amount of corrosion should weaken it significantly. But, make no mistake, converting a poor XJ-S into a good one is going to cost a lot of money, so it makes sense to buy a good one."

Newer models shouldn't suffer from the dreaded tinworm in the same way, but should still be given a thorough independent inspection.

Engine
The V12 engine is fairly bullet-proof, as it is so rarely under any kind of stress. The usual type of inspection is called for here: telltale signs of a blown head gasket, healthy oil pressure (around 60psi on a warm engine whilst running at speed is ideal), rattly timing chains and so on; oil consumption should be very low. Just check that the rear crank seal is dry, otherwise this could result in a hefty repair bill. The unit should be very quiet and very smooth, but, as always, if in doubt, have it looked over by a professional - it could save you a fortune.

The six-cylinder engines are also strong, but again regular servicing is the key. In the author's opinion, the AJ6 3.6 was not as nice as the 4.2 XK, but the later four-litre AJ16 version is a cracking unit. Again, the usual checks should suffice.

If the car is an older one, perhaps with a high mileage, treat it to regular services with shorter gaps than the handbook states. For the sake of a few gallons of oil and the odd filter, it's better to be safe than sorry, especially with a V12. If you want to keep up the service history, some Jaguar-Daimler dealerships are now offering superb servicing deals on older vehicles, and there are any number of specialists dotted around the country.

Transmission
GM automatic gearboxes have been pointed out as a weak spot in previous guides, but the author hasn't come across any major problems. The only thing that springs to mind was a noise on initial start up, but this disappeared after a filter change.

However, with either the GM or original Borg-Warner unit, it will certainly pay to check the colour and condition of the transmission oil - if it's brown or black, walk away, as a big bill is just around the corner; changes should be smooth and precise, with no sign of a slipping clutch. The ZF box was fitted on very late cars; these vehicles should come with a service history.

Manual transmissions have a good reliability record, so just the normal checks should suffice, but remember clutch assemblies do not come cheap, and if it is an early V12, was it fitted with a manual box originally? Engine and gearbox mountings will almost certainly need replacing on older cars: whilst underneath the vehicle, remember to look for oil leaks on the differential housing.

Suspension and steering
Watch for worn bottom balljoints and wheel bearings up front, and the condition of rubber bushes throughout the suspension - there are lots of them and they make an big difference to the way the car drives. To replace a few at a time doesn't hurt the bank balance too much, especially if you can do a lot of the work yourself, but be prepared for a big bill on a car bought for restoration. The front shock absorbers take a lot of weight and have to work very hard under braking; if you have to replace them, do so with matched pairs.

At the rear, there are more bushes to check, especially the rubber radius arm bushes. In fact, the radius arms themselves need to be looked at carefully for signs of corrosion, and subframe and axle mountings should also be inspected. Again, check the wheel bearings and universal joints. The car should have been greased regularly if unnecessary wear is to be avoided, so a proper service history is a real plus-point.

The power steering caused the author a lot of problems; even after the rack was replaced with a reconditioned unit, there was trouble with the hoses and then the pump itself. Steering rack bushes are a popular cause for concern, and many people replace them with a harder aftermarket alternative.

Braking system, wheels and tyres
Front discs often seem to warp, but replacement is simple enough and not too expensive; as with shock absorbers, always replace discs in matched pairs.

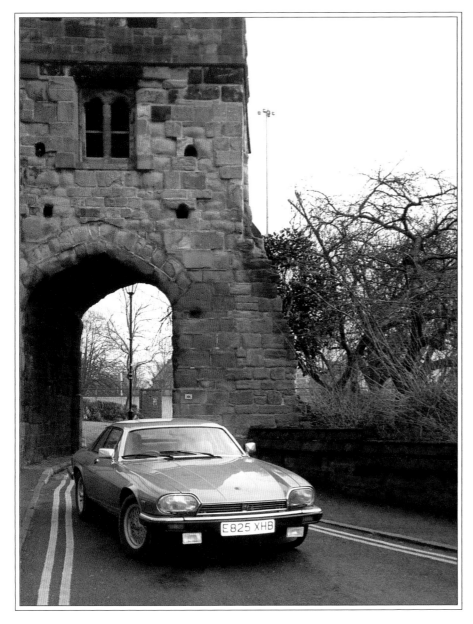

Ensure the air conditioning is working properly, as it could be quite costly to repair (it could just as easily be a simple service, but it's always best to assume the worst when you're the buyer). Electrics are perhaps the biggest concern, so check all electrical systems carefully: replacement lights can be quite expensive. Make sure the electric windows are working - as the author found out, the price of a new motor is truly frightening.

Modified parts

A large number of companies specialize in performance upgrades and body kits for the XJ-S. Some have been covered in the main text, but the list is virtually endless and growing daily. It would pay to subscribe to a quality magazine like Jaguar World that carries specialist advertising and details of all the latest products on offer.

Which one?

I have probably painted a very bleak picture in this chapter, but the XJ-S experience is just as good today as it has always been. In a recent edition of Your Classic, Robert Coucher wrote: "On the smooth Surrey dual carriageways the big Jag's in its element. The engine's inaudible, the steering fingertip-light, the feeling of serenity captivating. That special Jaguar trait; grace, space and pace, is there all right. Just lean on the throttle and nothing untoward happens except the horizon starts rushing towards you. The 5.3 litre engine's sublime, some say the best in the world.

"But the real surprise comes on

Replacing the rear disc brakes on older models is a real struggle, although pad replacement is fairly straightforward; the handbrake is not the best in the world, but should pass the MoT at the very least.

The various alloy wheels tend to go fairly grubby in time, as the clear protective coating comes away from the wheel and corrosion sets in from behind. Eventually, this coating breaks and the only remedy is to have the wheel restored at a specialist shop.

Tyres can be sourced quite cheaply

nowadays, but check for uniform wear patterns before discarding a worn set. As with all XJ models, tyres give a very good indication of suspension faults.

Trim and electrics

Interiors generally last well, but bear in mind that leather trim, although readily available, is expensive to replace; carpeting is also quite costly should the original be past its best. The wood on the HE models often peels away on the centre console, although replacement pieces are available.

the twisty bits. Throwing the V12 about is a revelation. No lurching or sliding, the Jaguar is great fun to hustle along, the double wishbone suspension soaking up crests and bumps with great aplomb."

An XJ-S, like most of the older Jaguars, can be run on a budget, but this is not recommended. The only way to run a car of this type is to keep throwing money at it, as, long-term, it will spend less time off the road and repair bills will be kept to a minimum. Bearing this in mind, buy what you can afford to run properly (i.e. don't spend your last penny to secure the vehicle), and join one of the many Jaguar clubs worldwide - the initial investment in membership will almost certainly pay for itself many times over.

The best advice is pay a little extra in the first place for the right car (preferably one with a good service history) and have it checked by an independent garage, the AA/RAC or someone familiar with the model.

Spares are still plentiful in the main, but some trim is now quite hard to find, and original TWR components are becoming very scarce (reconditioning is often the only path open for some mechanical parts). There are many companies dealing in spares, such as David Manners, who has vast stocks on the shelf, Ray Ingham at Classic Spares (who also races an XJ-S) - the list could go on forever. There are also a number of sources throughout the UK for secondhand parts.

Which S you choose is going to be down to personal preference. The closed coupé, cabriolet and convertible are all so different in character, and then there are the engine options to take into account. Add in the TWR models, early or later styling, and it becomes something of a minefield. Price will also be a factor for most of us.

If I had to choose just one model, it would be the 1988.5 model year V12 coupé. It retains the original styling with HE upgrades, but comes fully-loaded with a whole host of goodies fitted as standard, and represents excellent value for money on the secondhand market. As a long-term owner - and someone who has driven almost every one of the breed - I can confirm that this is a superb Grand Tourer which, unlike so many of its 'exotic' counterparts, is suitable for everyday use.

The XJ-S is never going to be cheap to run, but it really grows on you - it has to be one of the world's most elegant automobiles, especially in late-HE form. The profile, when looking down on the car from a front or rear three-quarter view, has to rank alongside any piece of sculpture found in an art museum.

APPENDIX II

PRODUCTION
FIGURES

These figures have been arranged to give yearly production totals for all XJ-S models, along with a breakdown for coupé, cabriolet and convertible production by engine type (the XJR-S, in all its forms, has been included in the relevant V12 column).

Year	V12 Coupé	6-cyl Coupé	V12 Cabrio.	6-cyl Cabrio.	V12 Conv.	6-cyl Conv.	Annual Prod.	Grand Total
1975	1245	-	-	-	-	-	1245	1245
1976	3082	-	-	-	-	-	3082	4327
1977	3890	-	-	-	-	-	3890	8217
1978	3121	-	-	-	-	-	3121	11,338
1979	2405	-	-	-	-	-	2405	13,743
1980	1057	-	-	-	-	-	1057	14,800
1981	1292	-	-	-	-	-	1292	16,092
1982	3088	18	-	5	-	-	3111	19,203
1983	4446	131	-	101	-	-	4678	23,881
1984	5852	451	7	199	-	-	6509	30,390
1985	6067	782	709	393	-	-	7951	38,341
1986	6641	650	1567	194	-	-	9052	47,393
1987	6699	1362	1510	196	59	-	9826	57,219
1988	5355	2066	132	-	2803	-	10,356	67,575
1989	4330	1999	-	-	4877	-	11,206	78,781
1990	3267	1326	-	-	4633	-	9226	88,007
1991	1507	1355	-	-	1400	387	4649	92,656
1992	553	1175	-	-	824	1042	3594	96,250
1993	507	1277	-	-	657	2716	5157	101,407
1994	242	985	-	-	1058	4633	6918	108,325
1995	239	852	-	-	309	4032	5432	113,757
1996	19	121	-	-	29	1487	1656	115,413

Total 12-cyl coupés:	64,904
Total 6-cyl coupés:	14,550
Total coupé production:	**79,454**
Of which, sales to USA were:	33,493
Total 12-cyl cabriolets:	3925
Total 6-cyl cabriolets:	1088
Total cabriolet production:	**5013**
Of which, sales to USA were:	1912
Total 12-cyl convertibles:	16,649
Total 6-cyl convertibles:	14,297
Total convertible production:	**30,946**
Of which, sales to USA were:	23,819*
Total XJ-S production:	**115,413**
Of which, sales to USA were:	59,224

* Figure includes H&E conversions.

INDEX

Jaguar Cars Ltd, its products and subsidiaries are mentioned throughout the book.

Visit Veloce on the Web - www.veloce.co.uk